Energopolitics

Energopolitics

Wind and Power in the Anthropocene   **Dominic Boyer**

Duke University Press   *Durham and London*   2019

Library of Congress Cataloging-in-Publication Data
Names: Boyer, Dominic, author.
Title: Energopolitics : wind and power in the Anthropocene / Dominic Boyer.
Other titles: Wind and power in the Anthropocene
Description: Durham : Duke University Press, 2019. |
Includes bibliographical references and index.
Identifiers: LCCN 2018047267 (print)
LCCN 2019000347 (ebook)
ISBN 9781478004394 (ebook)
ISBN 9781478003137 (hardcover : alk. paper)
ISBN 9781478003779 (pbk. : alk. paper)
Subjects: LCSH: Wind power—Research—Mexico—Tehuantepec, Isthmus of. |
Renewable energy sources—Mexico—Tehuantepec, Isthmus of. |
Renewable energy sources—Political aspects. |
Electric power production—Mexico—Tehuantepec, Isthmus of. |
Energy industries—Mexico—Tehuantepec, Isthmus of. | Energy
development—Political aspects. | Energy policy—International
cooperation. | Geology, Stratigraphic—Anthropocene.
Classification: LCC TJ820 (ebook) | LCC TJ820 .B694 2019 (print) |
DDC 333.790972/62—dc23
LC record available at https://lccn.loc.gov/2018047267

Cover art: Grid, La Ventosa, 2013. Photo by Dominic Boyer.

This title is freely available in an open access edition thanks to generous
support from the Fondren Library at Rice University.

For Cymene

Contents

Joint Preface to *Wind and Power in the Anthropocene*

A Dynamic Duo

Welcome to our duograph. You may be entering into the duograph through *Ecologics* or *Energopolitics*, but in each case, we invite you to engage both sides of this work. The duograph is a new and experimental form that needs your active engagement. But what is a duograph? you might rightly ask. A duograph consists of two single-authored ethnographies that draw from a shared fieldwork experience and the same archive of research material. As a textual form, the duograph emerged from our field research (2009–13) on the political and ecological dimensions of wind power development in Mexico's Isthmus of Tehuantepec. The idea evolved partly out of experimental interest and partly out of necessity. The two of us spent many long evenings debating the significance of one aspect or another of the research and gradually found ourselves setting out from the center of the project in different theoretical and thematic directions. The fieldwork itself was a joint enterprise from start to finish; every interview, every meeting, every protest, involved both of us. We originally expected that the writing would follow a similar path toward a coauthored monograph. But while coauthoring offers many opportunities to learn and grow through dialogue, it also involves many compromises and ultimately must resolve in a synthetic voice and direction. We wanted to do this differently.

We eventually realized how important it was to each of us that we be able to tell a different part of the immensely complex story unfolding in the isthmus. Cymene wanted to spotlight the salience of human-nonhuman relations in energy transition while Dominic wished to concentrate on unraveling the political complexity of wind power. We decided to experiment by elaborating our different analytics and interests in companion volumes that are meant to be read together. A working definition of the duograph would be a conversation between researchers that materializes in two texts, which do not require analytic synthesis or consensus. We view the duographic form as a way to produce collaborative scholarship that helps to make visible the multiplicity of stakes and attentions existing within the practice of research collaboration. The observations and arguments found in each of these volumes emerged from close dialogue and are by no means incommensurable, but neither are they serial parts of the same narrative. They speak in parallel, but not always in unison. Characters, dynamics, and events crisscross them, but they are approached through different analytic lenses. We hope that the duograph offers an experimental prototype in collective authorship that may be of value to other collaborators and other projects elsewhere.

Wind Power in Mexico

Our ethnography addresses a central question of our anthropocenic times: How can low-carbon energy transition happen? Or, put differently, What happens in those transitions? Who sets the agenda? Who—human and otherwise—is affected? And what are the political (in the broadest sense of the term) forces that shape the possibilities for low-carbon energy futures?

These questions initially took shape at Busboys & Poets café in Washington, DC, in late 2008 as we prepared for a move to Houston, Texas, a global epicenter of the fossil fuel industry. We considered a number of different fieldsites of renewable energy production that appeared to be poised for rapid development. We looked at the DESERTEC solar project in Morocco and nascent programs of wind development in Venezuela and Brazil among other cases. But the one that attracted and held our attention most strongly was Oaxaca's Isthmus of Tehuantepec.

A gap in the Sierra Madre Mountains creates a barometric pressure differential between the Gulf of Mexico and the Pacific Ocean, forming a wind tunnel in the isthmus where wind speeds regularly flirt with tropical storm strength. The *istmeño* wind is capable of overturning semitrailers with ease, uprooting trees, and stripping the paint off boats. This region—often said to be the least developed in a state that is the second poorest in Mexico—is considered to have among the best resources for terrestrial wind power anywhere in the world. That potential was first tapped in the mid-1990s through government demonstration projects designed to lure transnational investment in renewable energy production. But wind development only really gained attention and momentum during the administration of President Felipe Calderón (2006–12). Although Calderón's administration is better known for its drug war and for ceding sovereignty to cartels and capital, his climate change advocacy transformed Mexico from a pure petrostate into a global leader in low-carbon energy transition. Mexico passed some of the most ambitious, binding clean-energy legislation anywhere in the world, including a legal mandate that 35 percent of electricity be produced from non-fossil-fuel sources by 2024, with 50 percent of that green electricity expected to come from wind power, and with most of that wind power expected to come from the Isthmus of Tehuantepec. Private-Public Partnerships (PPPS) in wind energy development mushroomed rapidly. Between 2008 and 2016 the wind energy infrastructure of the isthmus expanded from two wind parks offering 85 megawatts of production capacity to twenty-nine wind parks with 2,360 megawatts of capacity, a 2,676 percent increase in less than a decade that has made the isthmus the densest concentration of onshore wind parks anywhere in the world.

Over the course of sixteen months of field research (in 2009, 2011, and 2012–13), we sought to cast as broad a net as possible and speak with representatives of every group of "stakeholders" in wind development in Mexico. Conversations with community members and corporate executives; federal, state, and local government officials and NGO staff; industry lobbyists and antiwind activists; conservationists and media professionals; indigenous rights advocates, bankers, and federal judges, all provided a meshwork of perspectives, which we traced as we moved between the many communities of the isthmus; to the state capital, Oaxaca City; and finally to the federal capital, Mexico City. In total, we conducted more than three hundred interviews and participated in hundreds of hours of less formal conversations. Working with a team of local researchers, we were able to conduct the first door-to-door survey of reactions to wind development in La Ventosa—one of two isthmus towns that are now nearly completely encircled by wind parks. We sat in on governmental and activist strategy meetings and toured wind parks. We marched, rallied, and stood at the fulcrum of many roadblocks erected by opponents of the wind parks. We witnessed the evolving politics of solidarity between *binnizá* (Zapotec) and *ikojts* (Huave) peoples whose shared resistance to particular forms of energy infrastructure brought them into alliance after hundreds of years of interethnic conflict. We arrived at and left fieldwork as committed advocates for low-carbon energy transition. But our experiences in Mexico taught us that renewable energy can be installed in ways that do little to challenge the extractive logics that have undergirded the mining and fossil fuel industries. Renewable energy matters, but it matters more how it is brought into being and what forms of consultation and cooperation are used. We thus came to doubt that "wind power" has a singular form or meaning. Everywhere in our research, it was a different ensemble of force, matter, and desire; it seemed inherently multiple and turbulent, involving both humans and nonhumans. To capture that multiplicity, we came to think about our object of research as "aeolian politics," borrowing from the Spanish term for electricity derived from wind power, *energía eólica*.

Three case studies of aeolian politics came to absorb us in particular—Mareña Renovables, Yansa-Ixtepec, and La Ventosa—the first is the most complex and is treated at length in the *Ecologics* volume. The other two are highlighted in the *Energopolitics* volume. All three represent distinct configurations of aeolian politics; two can be categorized as cautionary tales of failure and the other as an example of the successful achievement of what for many is the renewable dream come to life. And yet success and failure were always in the eyes of their beholders. In all three studies we have sought to balance the fact of anthropogenic climate change and the need for global decarbonization against the local salience of vulnerable statecraft, demands for indigenous sovereignty, and the other-than-human lives that inhabit the Isthmus of Tehuantepec.

Volumes

ECOLOGICS

Ecologics tells the story of an antidote to the Anthropocene, one that was both a failure and a success. The Mareña Renovables wind park would have been the largest of its kind in all Latin America, and it promised immense reductions in greenhouse gas emissions as well as opportunities for local development. In *Ecologics* we follow the project's aspirational origins as well as the conflicts and ethical breakdowns that would leave it in suspension. Drawing from feminist theory, new materialisms, and more-than-human analytics, this volume of the duograph examines the ways that energy transitions are ambivalent: both anticipatory and unknown, where hope and caution are equally gathered. In the case of Mareña Renovables, distinct imaginaries of environmental care and environmental harm were in conflict, effectively diagnosing the deeply relational qualities of energy and environment. The core argument that *Ecologics* advances is that the contemporary dynamics of energy and environment cannot be captured without understanding how human aspirations for energy articulate with or against nonhuman beings, technomaterial objects, and the geophysical forces that are at the center of wind power and, ultimately, at the heart of the Anthropocene.

The analytic architecture of *Ecologics* is both anticipatory and interruptive, and readers are encouraged to engage with the work in an itinerant and wandering way. Three chapters focus on the case of the Mareña project, tracing its inception and the policy regimes and economic conditions that allowed for its initial development (chapter 2, "Wind Power, Anticipated"), following it through a series of dramatic standoffs and protests against the park's creation among indigenous and mestizo communities in the isthmus (chapter 4, "Wind Power, Interrupted"), and finally witnessing the collapse of the wind power project itself resulting from multiple political, economic, and communicational impasses (chapter 6, "Wind Power, In Suspension"). These chapters are interrupted by others that focus on wind, trucks, and species respectively. The interruptive design is intended to mime the empirical, ethnographic dynamics of the research, where forces (like wind), technomaterial tools (like trucks), and other-than-human beings (creatures of all kinds) came to stall and vex human-designed notions of progress and infrastructural development. In *Ecologics* creatures, materials, and elemental forces are bound up with wind power as an analytic object, and they in turn invite new human responses to the paradoxes we face in a time of climatological uncertainty.

ENERGOPOLITICS

Energopolitics engages the case of Mexican wind power to develop an anthropological theory of political power for use in the Anthropocene anchored by discussions of "capital," "biopower," and Dominic's own neologism, "energopower." At the same time, the volume emphasizes the analytic limitations of these conceptual minima when confronted with the epistemic maxima of a situation of anthropological field research on political power. Those maxima not only exceed the explanatory potential of any given conceptual framework, they also resolutely demand the supplementary analytic work of history and ethnography. Concretely, the volume argues that to understand the contemporary aeolian politics of the Isthmus of Tehuantepec, one needs to understand, among other things, a contested history of land tenure, *caciquismo* (boss politics), and student/teacher/peasant/worker/fisher opposition movements specific to the region; the phantasmatic status of state sovereignty within Mexican federalism; the clientelist networks and corporatist machinations of the Mexican political parties; the legacies of settler colonialism; a federal government anxious about waning petropower and climate change; and a vulnerable parastatal electricity utility trying to secure its future in an era of "energy reform." These forces are just as critical to Mexico's aeolian politics as the processes and dynamics that are duly captured by concepts such as capital, biopower, and energopower. *Energopolitics* is thus an urgent invitation for Anthropocene political theory to unmake and remake itself through the process of fieldwork and ethnographic reflection.

The invitation unfolds across five ethnographic chapters, each highlighting a different localization of aeolian politics. We begin with the as-yet failed effort to build a community-owned wind park in Ixtepec, then move east to the town of La Ventosa, which is successfully encircled by turbines that were built in the dominant PPP paradigm, yet has also been beset by uncertainty and unrest. We encounter the performative sovereignty of the state government in Oaxaca City as it searches for a means to regulate and profit from wind development and then journey northwest to Mexico City to interview those in government, industry, and finance who firmly believe they are steering the course of wind power in the isthmus. Finally, we return to Juchitán, which is not only the hub of local aeolian politics in the isthmus but also a town whose citizens imagine themselves to be the inheritors of a decades- if not centuries-long tradition of resistance against the Oaxacan and Mexican states. In this way, *Energopolitics* seeks to speak *terroir* to *pouvoir*, highlighting the need to resist anthropocenic universalism by paying attention to the profound locality of powers, agents, and concepts. As Claire Colebrook has argued, recognition of the Anthropocene should mark the "return of difference" that has been long called for in feminist and ecological criticism.

Collaboration in Anthropology

Our duograph belongs to a long history of anthropological collaboration in research and writing. In the early decades of North American and European ethnology, the discipline's close ties to fields like geography and natural history meant that the scientific expedition was an important apparatus of anthropological research practice. In the late nineteenth and early twentieth centuries, projects of linguistic and cultural salvage and analysis remained closely allied with archaeology and museology, which explains how some of the most ambitious and important collaborative anthropological enterprises of the era—Franz Boas's Jesup North Pacific Expedition (1897–1902), for example—were organized principally around building natural history collections. As the twentieth century wore on, an individualistic model of field research came to predominate in American and European anthropology, at least normatively, and was celebrated for the transformative qualities of participant-observational immersion. But one would scarcely have had to scratch the surface of any ethnographer-informant dyad to illuminate the complex webs of social enablement—involving research assistants, translators, laborers, intermediaries, government agents—that made anthropological research in the classic Malinowskian mode possible.

After the Second World War, a new emphasis on interdisciplinary area studies research in the social sciences expanded and intensified anthropology's range of collaborative engagements around the world. Much as expedition-era anthropology was absorbed into colonial and imperial knowledge projects, the area studies era was imbricated with the national and international political dynamics of the Cold War. Governments sought to enroll anthropologists in military and intelligence operations across the world—Project Camelot being one of the most well known. However, anthropology was also broadening its epistemic ambitions and moving from cultural salvage projects toward a grappling with modernity and the complex cultural and social dynamics of cities, nations, and world systems. Interdisciplinary exchanges no doubt served to accelerate this shift. And 1950s enterprises like Cornell's Vicos project in Peru (creating a "laboratory for social change") or the MIT Modjokuto project in Indonesia (which gave Clifford and Hildred Geertz their first fieldwork opportunity) cultivated the kinds of long-term interdisciplinary research networks that influenced graduate training and pedagogy as well.[1]

The postwar period also saw an efflorescence of anthropological research partnerships mediated through marriage and other life partnerships. Margaret Mead and Gregory Bateson are a classic example, Margaret Mead and Ruth Benedict a more elusive but possibly more substantial one. Then came the Geertzes as well as June and Manning Nash, Marilyn and Andrew Strathern, Edith and Victor Turner, and Margery and Eric Wolf, followed later by Barbara and Dennis Tedlock, Michelle and Renato Rosaldo, Sally and Richard Price, and Jean and John Comaroff, among others. Anthropology has seen many couples practice the crafts of research, teaching, and writing under at least a partly shared sense of identity, each navigating its own relational dynamics as well as the dominant masculinist heteronorms of the discipline and the university in the twentieth century.

Reacting to the still broader and more complex scale of post-1980s globalization and its social, economic, and environmental consequences, the twenty-first century has seen renewed interest in collaborative research partnerships. Three that have inspired our duograph in particular have been the Matsutake Worlds Research Group (Anna Tsing, Shiho Satsuka, Miyako Inoue, Michael Hathaway, Lieba Faier, and Timothy Choy), the Ethnographic Terminalia collective (Craig Campbell, Kate Hennessy, Fiona McDonald, Trudi Lynn Smith, and Stephanie Takaragawa), and the Anthropology of the World Trade Organization group (Marc Abélès, Máximo Badaró, Linda Dematteo, Paul Dima Ehongo, Jae Aileen Chung, Cai Hua, George Marcus, Mariella Pandolfi, and Phillip Rousseau).[2] All are multi-institutional and international partnerships that have explored new ways of creating anthropological knowledge by crossing the boundaries between anthropological research practices and the arts.

Collaboration itself is nothing new in anthropology; there is abundant evidence that it has been a productive dimension of anthropological research and writing since the discipline's beginning. Further, intimate research partnerships have long fueled the production of anthropological knowledge. There is doubtless an important book to be written about how the particular qualities, subjectivities, and dynamics of particular collaborations have influenced the kinds of knowing and knowledge that those enterprises generated. But our intervention here is more limited. We have found it striking that the spirit of collaborative research has not always translated well into practices of authorship. Coauthored texts remain the exception rather than the rule in anthropology, even when they derive from jointly undertaken field research.[3] The reasons for this gap are not simple and involve considerations ranging from professional reputation to relational dynamics to institutional audit cultures that seek to impose a mathematics of individual accomplishment and accountability on the sociality of research, analytic, and writing practices. What is striking in our view is that there are relatively few models for collaborative writing beyond the model of the jointly authored single text that synthesizes analytic perspectives under a common "we." This is why we have centered our methodological intervention on the duographic form: we are looking for ways to strike a better balance between individual ideation and expression and collaborative fieldwork and archiving.

An important added benefit of the duograph is that it permits a more extensive analytic division of labor between its volumes, as parallel yet distinct arguments can be developed with respect to the common research archive. In our case, the *Ecologics* volume's close focus on how human energetic and environmental aspirations intersect with other-than-human beings and agencies complements yet also reframes the *Energopolitics* volume's effort to offer a more nuanced and comprehensive set of analytics of (human) political power, and vice versa. If the general premise of the entire research project has been that a certain politics of energy is creating a situation of ecological emergency, then it is fitting, and we might say necessary, to be able to offer detailed conceptual and ethnographic accounts of both sides of the equation—energopolitical and ecological. Had we tried to compact all these storylines into a single, synthetic account, however, we might well have burst its seams or have been forced to simplify matters to the extent that neither side would have received its due. In the duographic form, meanwhile, two volumes working together in the mode of "collaborative analytics" can dive deeply into different dimensions of the research while still providing valuable ethnographic elaboration and conceptual infrastructure for each other.[4]

Your Turn

One of our favorite rationales for the duograph is what is happening right now: you are deciding where to start. True to the lateral media infrastructures and expectations of this era, we aspire to offer a more dialogic, collaborative matrix of encounter with anthropological writing. We have sought the words to write; you now seek the words to read. We have left signposts as to where we think the volumes intersect. But you can explore the duograph as you like, settling into the groove of one narrative or zigzagging between them. Think of it somewhere between a Choose Your Own Adventure book and open-world gameplay. Follow a character, human or otherwise; riddle through the knots and vectors of aeolian politics; get bogged down somewhere, maybe in the politics of land or the meaning of trucks; then zoom back out to think about the Anthropocene. Or perhaps pause for a minute or two to watch the birds and bats and turbines that now populate the istmeño sky.

Cymene Howe and *Dominic Boyer*

Acknowledgments

We gratefully acknowledge the following persons and institutions whose support made this duograph possible. The preliminary field research for the project was funded by a grant from the Social Sciences Research Institute at Rice University. Rachel Petersen and Briceidee Torres Cantú offered extraordinary research assistance in the early research phase. The main period of field research in 2012–13 was funded by the Cultural Anthropology Program of the Division of Behavioral and Cognitive Sciences at NSF (research grant #1127246). We further thank the Social Sciences Dean's Office and especially our colleagues in Rice Anthropology—Andrea Ballestero, James Faubion, Jeff Fleisher, Nia Georges, Susan McIntosh, and Zoë Wool—for absorbing our share of administrative, advising, and teaching duties to allow us to undertake field research for a full year. We give thanks as well to our dear colleagues in Rice's Center for Energy and Environmental Research in the Human Sciences (CENHS)—Bill Arnold, Gwen Bradford, Joe Campana, Niki Clements, Farès el-Dahdah, Jim Elliott, Melinda Fagan, Randal Hall, Lacy M. Johnson, Richard R. Johnson, Jeff Kripal, Caroline Levander, Elizabeth Long, Carrie Masiello, Tim Morton, Kirsten Ostherr, Albert Pope, Alexander Regier—and the Center for the Study of Women, Gender and Sexuality (CSWGS)—Krista Comer, Rosemary Hennessy, Susan Lurie, Helena Michie, Brian Riedel, Elora Shehabuddin, and Diana Strassmann—for much-appreciated moral and intellectual support during the writing phase of the project.

In the Isthmus of Tehuantepec, our debts are many. We would first of all like to thank the Istmeño binnizá and ikojts communities of Álvaro Obregón,

Ixtepec, Juchitán, La Ventosa, San Dionisio del Mar, and Unión Hidalgo for hosting us during different phases of the field research and for a great many conversations and small acts of kindness that helped this project immeasurably. There were certain individuals in the isthmus who became friends and allies in the course of this research and without whose support this duograph would never have been possible. Our special thanks to Isaul Celaya, Bettina Cruz Velázquez, Daniel González, Mariano López Gómez, Alejandro López López, Melanie McComsey, Sergio Oceransky, Rodrigo Peñaloza, Rusvel Rasgado López, Faustino Romo, Victor Téran, Vicente Vásquez, and all of their families. Our condolences to the family of David Henestrosa, a brave journalist who passed away shortly after our field research was completed.

In Oaxaca City, Mexico City, and Juchitán, there were several organizations whose representatives helped enable particular parts of our research, including AMDEE, APIIDTT, APPJ, CFE, CNTE (Sec. 22), CRE, EDF Energies Nouvelles, Iberdrola Mexico, SEGEGO, SEMARNAT, SENER, UCIZONI, and the Universidad Tecnológica de los Valles Centrales.

We would also like to thank several individuals who helped us navigate the complex world of Mexican wind power. Thanks to Fernando Mimiaga Sosa and his sons for convincing us to make Oaxaca the focus of our research, to Sinaí Casillas Cano for his insights into the political contingencies of Oaxacan state governance of renewable energy development, to Felipe Calderón for answering our questions concerning the importance of wind power during his sexenio, to Laurence Iliff for a wealth of insights into the behind-the-scenes politics of Mexican energy, to Dr. Alejandro Peraza for an invaluable series of contacts among the power brokers of Mexican energy, and to The Banker, who remains anonymous but who gave us critical insights into the financial dimension of wind power development.

Both the project and the duograph have benefited immensely from comments at several meetings of the American Anthropological Association, the Cultures of Energy Symposium Series, and the Petrocultures Research Group. Conversations with colleagues in the following forums were also critical in shaping the thinking that has gone into the manuscripts: the Climate Futures Initiative, Princeton University; Association for Social Anthropology Annual Conference, Durham, England; Earth Itself: Atmospheres Conference, Brown University; Energy Ethics Conference, St. Andrews University; Religion, Environment and Social Conflict in Contemporary Latin America, sponsored by the Luce Foundation and the Center for Latin American and Latino Studies, American University; Hennebach Program in the Humanities, Colorado School of Mines; Klopsteg Seminar Series in Science and Human

Culture, Northwestern University; Colloquium for the Program in Agrarian Studies, Yale University; Rice University's Feminist Research Group and CENHS Social Analytics workshop; Environments and Societies Institute, University of California, Davis; Global Energies Symposium, University of Chicago Center for International Studies; Durham Energy Institute; Yale Climate and Energy Institute; Rice University, Mellon-Sawyer Seminar; Centre for Research on Socio-Cultural Change, Manchester University; After Oil, University of Alberta, Edmonton; Institut für Europaïsche Ethnologie, Humboldt Universität zu Berlin; Electrifying Anthropology Conference, Durham University, sponsored by the Wenner-Gren Foundation; Parrhesia Masterclass, Cambridge University; Alien Energy Symposium, IT University of Copenhagen. Thanks also goes out to friends and colleagues at the Departments of Anthropology at the London School of Economics; Ustinov College, Durham University; University of California, Los Angeles; University of California, Davis; Yale University; and University College London who were kind enough to invite us to share our work-in-progress with them.

Paul Liffman has been immensely helpful throughout the life of this project, offering smart critique and always ready to provide context by drawing on his truly encyclopedic knowledge of Mexico's political, economic, and social worlds. Our collaboration with Edith Barrera at the Universidad del Mar enriched our field experience, and back in the United States we were lucky to work with Fiona McDonald, Trudi Smith, and Craig Campbell, members of the Ethnographic Terminalia Collective, to imagine/incarnate the wind house installation of Aeolian Politics at the Emmanuel Gallery in Denver, Colorado. Lenore Manderson was also kind enough to review that project of imagining ethnography and wind differently. Anna Tsing and John Hartigan provided marvelous feedback on earlier chapters, and in later stages, the duograph benefited greatly from anonymous reviewers' incisive commentaries. We appreciate especially their willingness to think with us on the experimental form of the duograph and for their wise handling of interconnections between the volumes.

Thanks are also owed to the wonderful professionals at the University of Chicago Press, especially Priya Nelson, as well as Stephanie Gomez Menzies, Olivia Polk, Ken Wissoker, Christi Stanforth, and Courtney Baker at Duke University Press.

Introduction

Anthropolitics Meets Anthropology in the Anthropocene

This volume of the duograph is concerned with the relationship of energy to power in the context of the Anthropocene. It seeks to highlight and explore how the material and infrastructural dimensions of energy both enable and disable certain configurations of political power. The line of analysis questions whether political power in the conventional (human-centered) sense can really be taken to be an autonomous and efficacious domain. This has both theoretical and practical implications at a time when political power is a growing "matter of concern" in the struggle against processes like global warming and species extinction, when politicians are called upon to "get serious" about climate change, and when governments are implored to plan for energy transition or resilience to rising sea levels.[1] How can we reform a human-centered understanding and practice of politics—*anthropolitics*—so that it can adequately comprehend and address the conditions and challenges of the Anthropocene?

When Cymene and I wrote the proposal to the National Science Foundation for the grant that would eventually fund the main period of our field research, we more or less took for granted the significance of human political power in addressing climate change. We said we wished to investigate the "political culture" of wind power development in southern Mexico in order to understand how a "vulnerable state" like Mexico was going to be able to orchestrate a diverse and potentially contentious field of stakeholders and follow through on the federal government's ambitious clean electricity production targets. We questioned whether "states, especially those already struggling to meet their current governmental obligations, possess the political authority to implement important programs of national development such as renewable energy." But we did not question whether "political culture" itself—a term we used in a deliberately expansive way to signal not only the interactions between states and citizens but also the political negotiations and exchanges among stakeholders including local landowners, activists, political parties, NGOs, journalists, and representatives of transnational corporations—was an assemblage from which one might reasonably expect efficacious responses to climate change and strategies for energy transition to emerge.

But is the efficacy of political culture in the Anthropocene actually a reasonable assumption? What we did *not* emphasize in our grant proposal—and this is very likely because we were appealing to another equally anxious, if not equally vulnerable, state for funding—is that one of the articles of faith of the past thirty years of global governance is that market and technological forces are better positioned to answer any dilemma than any existing or imaginable configuration of political actors, instruments, and imaginations. The incremental and perpetually disappointing series of COP (Conference of Parties) meetings—even including the comparatively successful Paris COP 21 meeting in December 2015—have only seemed to reinforce this sense that the political process is inadequate to the task of engaging problems as massive and time sensitive as global warming.[2] Across the world, liberal political institutions seem too compromised by corporatist and populist influence, too much in the hands of political professionals operating according to their own temporalities and interests, and too belabored by inflexible, archaic political technologies to compete with entrepreneurs and engineers when it comes to delivering solutions. This skepticism has been paralleled in the discursive realm of political theory as figures ranging from Wendy Brown to Chantal Mouffe to Jacques Rancière to Peter Sloterdijk to Slavoj Žižek have diagnosed the economization, overformalization, militarization, spectacularization, and technicization of liberal political institutions that have led to a golden age of political ritual, cynicism, and theater at the expense of a capacity for a literal politics that might be able to address urgent anthropocenic processes of common (and not only human) concern such as droughts, flooding, warming, desertification, species extinction, plasticization, and oceanic acidification.[3] Probing more deeply into the scar tissue of late liberalism, Elizabeth Povinelli offers a reminder that "biontological" crises are nothing new on the frontiers of settler late liberalism, and she recommends that we take the wisdom of those who live on those frontiers more seriously. In the register of a probative and experimental "Karrabing analytics," she writes, "The earth is not dying. But the earth may be turning away from certain forms of existence."[4] Featuring prominently among those dying forms of existence is what I have termed "androleukoheteropetromodernity," an ugly word commensurate with the ugly lifeworld designed over the past several centuries to enable the dominion and luxury of hypersubject white men.

Yet back in the political centers of late liberalism, such a reckoning is held at bay by affective and epistemic investments in "markets" and "technology" as generative nexuses of innovations and solutions. These investments are also intimately attached to the political apparatus we call "neoliberalism." Whether one wishes to schematize that apparatus through the unfolding dynamics of capital and its class affiliations or through the evolving relay networks of power knowledge, it is obvious that neoliberalism has had a historicity unto itself; that is to say, it came into being, rose to global authority, and is now—so it seems anyway—in a state of gradual dissolution.[5] As that dissolution spreads, one finds that new and heterogeneous political potentialities are emerging. Whether one thinks of the racialized authoritarian movements rapidly gaining ground in many parts of the world today or the Arab Spring, the Occupy movement, the *indignados* of Spain, or the indigenous *asambleas* that are coming into being around wind parks in the Isthmus of Tehuantepec, it is difficult not to feel that new politics are seeking to be born that do not wish to be constrained by the inherited -isms (e.g., liberalism, socialism, communism, fascism, anarchism) of nineteenth-century European political philosophy. Neoliberalism appears to have lost much of its credibility and vitality as a world-making political ideology over the past decade. And yet it has no obvious heir-in-waiting, especially at the global level. Instead, a multitude of political experiments are emerging, often investing political attention and energy into smaller spheres of action. Some of these experiments, it goes without saying, embrace oppression, exclusion, and hatred in the manner of "integralist" movements past;[6] some perhaps augur a "time of monsters."[7] On the other hand, one finds movements committed to peace and humanism in unprecedented ways—Iceland's Best Party for one example.[8] Measured within this human's lifespan, there has never been a more invigorating time to think about political power. Things are *happening* in the world of politics, but they are escaping our conventional categories of analysis, often reducing analysts to a stilted language of "neos" and "posts" that, at the end of the day, seem inadequate for comprehending the processes of political formation we are witnessing.

Although perhaps an accidental conjuncture, the dissolution of neoliberal authority has occurred more or less simultaneously with wider mediatization and recognition of the Anthropocene. This has meant that there has been a vibrant zone of political experimentation focusing specifically on remediating anthropocenic vectors—for example, Transition Culture and the degrowth movement.[9] It has also stimulated some critics of neoliberal capitalism—Naomi Klein and David Graeber among others—to argue that acknowledgement of the Anthropocene marks a definitive beginning of the end for capitalist consumerist society since we have finally come to experience the deterioration of ecological systems at a planetary level.[10] The 1970s "limits to growth" and "tragedy of the commons" debates have been reactivated. But other theorists have abandoned the anthropolitical in favor of the ecopolitical. As Claire Colebrook writes, "The Anthropocene seems to arrive just as a whole new series of materialisms, vitalisms, realisms, and inhuman turns require 'us' to think about what has definite and forceful existence regardless of our sense of world."[11] Although it is possible to take the antianthropocentric turn in the human sciences as another reminder of why paying attention to human political power seems quaint in the contemporary world, I would rather take it as a challenge and opportunity to recalibrate the anthropolitical to a postanthropocentric conceptual universe. In other words, let us ask, How, where, and to what extent does and should human political power matter in the contemporary world? I have argued elsewhere that human agency, at a planetary scale, is difficult to deny given not only the various phenomena clustered under the Anthropocene rubric but also the generative potentialities of practices like synthetic biology and nanoengineering on the one hand and the destructive potentialities of advanced weaponry on the other.[12] Even if we have truly never been modern in an ontological sense, the fact that some humans have been behaving as though they were modern for centuries now—creating a potent instrumentarium for terraforming/anthroforming the planet for their convenience along the way—demands accountability and, one hopes, remediation. This means, I would argue, that an interest in understanding or influencing the anthropolitical cannot be bundled together with a rejection of anthropocentrism, fair though that rejection may be.

And here I would modestly propose that anthropology has an important role to play. As one of the more reflexively oriented disciplines within the human sciences, at least since the 1970s,[13] anthropology has viewed its own methods and objects of investigation with no small amount of skepticism. However, as the one discipline that has anthropos inscribed in its very jurisdiction, it seems unlikely that anthropology will ever fully commit itself to a posthuman turn. As such, anthropology seems an excellent "culture of expertise" from which to stage a reproportionalization of the human in the human sciences. Moreover, whatever soaring theoretical, even philosophical, aspirations anthropological knowledge might have, these are always connected umbilically to the sociality and materiality of a changing world of humans and nonhumans.[14] Anthropological knowledge is perpetually incomplete, disrupted, uncertain, somehow less than the sum of its parts. It is the right kind of knowledge for grappling with what Anna Tsing and her collaborators have termed "a damaged planet."[15] If Isabelle Stengers poses a cosmopolitical question, "How, by which artifacts, which procedures, can we slow down political ecology, bestow efficacy on the murmurings of the idiot?" then I would argue that the murmurings of the idiot—meaning not a fool but one who provides what Claude Lévi-Strauss once termed "the other message"—is precisely the domain of anthropological knowledge.[16] It is thus an apt domain from which to elicit "hyposubjectivity" in the face of Timothy Morton's diagnostics of "hyperobjectivity."[17]

WITH THESE PRELIMINARIES in mind, a reader solely interested in the ethnography of wind power in the Isthmus of Tehuantepec could skip ahead to chapter 1. The remainder of the introduction is a more detailed discussion of theory of power in the Anthropocene. My purpose is twofold. On the one hand I wish to schematize the conceptual minima of an anthropological theory of political power for use in the Anthropocene. Concepts such as capital and biopower clearly belong to those minima, not least because of the profound influence that Marxist and Foucauldian theory have exerted over human-scientific analysis of political power in the past several decades. And, while their status is more contested, I think it is, for the same reason, necessary to briefly discuss the psychoanalytic theory of desire and Brian Massumi's "ontopower" for what they might contribute to the analytics of political power. Since all these concepts share an inattentiveness to the energo-material contributions of fuel and electricity to political power, I also put forward my own neologism, "energopower," to expand the set of minima in a direction that I believe is analytically crucial for understanding our (the planet's, our species') contemporary conditions. On the other hand, I wish to emphasize the delicacy, one might even say the preciousness, of these conceptual minima when confronted with the epistemic maxima of a situation of anthropological field research on political power. Those maxima not only exceed the explanatory potential of any given conceptual framework, they also resolutely demand the supplementary analytic work of history and ethnography. For example, to understand the contemporary political culture of wind power development in the Isthmus of Tehuantepec, one needs to understand a deep history of colonial resource extraction and the more recent politics of land tenure and *caciquismo* (boss politics) specific to that region as much as one needs to understand the processes and dynamics captured by concepts such as capital, biopower, and energopower.

To an anthropologist this may simply appear to be a sky-is-blue statement about the importance of ethnography to complement and validate theoretical intervention.[18] But, as suggested above, my broader purpose here is to argue that if political theory wants to get serious about a critical engagement with anthropocenic phenomena, then it is going to have to become more anthropological along the way. The obvious problem baked into terms such as "Anthropocene" is their species-level universalism. This universalism often seems to be a rhetorical strength, especially when deployed to convince a species whose collective behavior is generating planetary effects while lacking a species-level political apparatus to take collective action. But without doubting that universalizing rhetoric can be efficacious in some contexts, the hailing of humanity as a species unfortunately obscures the differential culpability for global warming and environmental toxification, ignoring the fact that Northern empire has perpetrated these and other global conditions of precarity for centuries with impunity.[19] Univeralist rhetorics also typically obscure the fact that the reasons for anthropocenic action and the impasses to its recognition and transformation are highly, one might even say fundamentally, various. The reasons that one wind park near Cape Cod might be challenged in court and that another would be blockaded near Juchitán cannot be reduced to a general condition of NIMBYist self-interest. The same could be said of the support for utility-scale renewable energy projects in some corners of the global North and the support for postgrid energy solutions and degrowth in others. Moreover, the postanthropocenic futures that are being imagined and aspired to are equivalently multiple. For some, the struggle against the Anthropocene quests more or less to preserve familiar circuits of fulfillment in climatological terra incognita (e.g., sustainable green capitalism); for others, it is about resuscitating idealized past forms of life (e.g., nationalist nostalgia or indigenist restoration); yet for still others, it about a radical break with the past and present to prepare the way for hitherto-unrealized socio-ecological relations.

My argument, then, is that political theory needs to embrace the fact that, as Claire Colebrook has put it, "the Anthropocene is the return of difference."[20] Taking difference seriously means a willingness to think across scales, to recalibrate the capacity of "the local"—meaning both locus/place and also those beings who inhabit particular localities—to affect and transform the translocal.[21] It is in this respect that anthropological analysis of political power can play a valuable supplemental role as well as in Derrida's sense of supplement as that which reveals an originary lack.[22] Anthropological analysis thrives on the interillumination of translocal and local epistemics, on showing what universalizing schemata can and cannot reveal when confronted with an actual world of fluctuating, heterogeneous, and not infrequently contradictory signals.[23] *This is, then, a call for political theory to not so much "take ethnography seriously" as to accept ethnography's invitation to unmake and remake itself through the process of fieldwork.* Ethnography is a representational medium and as such will always game with words; those games do sometimes influence understanding for some interlocutors but not in the way that fieldwork, as ontomedium, can more fundamentally challenge and transform horizons and ways of knowing. If we wish to appreciate difference within the Anthropocene, fieldwork is a much-needed supplement to any theory of power.

As a proof of concept, this volume turns loose a certain set of power concepts in Mexico to show where they can help us to gain interpretive traction on specific events and dynamics and also where a variety of local forces and forms exceed or disable them.

Conceptual Minima: Capital, Biopower, Energopower

I have already proposed that capital, biopower, and energopower belong to the conceptual minima of an operational theory of political power in the Anthropocene. I discuss each of these terms in more detail below, but let me say at the outset that is this not intended to be a closed set of concepts, nor is my argument ontological in any sense. That is to say, first, there are many potentially valuable concepts missing here. I have selected these three not only to reflect key touchstones in recent political anthropological debate concerning the Anthropocene but also in light of the specificities of our case studies. Second, I resist (strongly) the idea that categorizing a type of power to enable the possibility of recognition and discussion corresponds to an argument for the being of power.[24] Capital, biopower, and energopower are conceptual lenses that help to bring into focus certain force relations. They do not represent singular forms of power per se. The "beings" that concepts like capital, biopower, and energopower signal should be regarded as multiplicities, diverse forces that have been bundled into more limited nominal forms as part of an analytical project.

Nonetheless, I would argue that none of these concepts can be functionally derived from the others, nor should any of the force clusters be trivialized with regard to the others. Each concept has its own analytic attentions and, by extension, its own uses. Karl Marx was able to use "capital" as an analytic instrument for examining the formalization, expropriation, and circulation of human productive activity, just as Michel Foucault was able to utilize "biopower" to explore objectification of and intervention in life from ethics to administration to science. As I have detailed elsewhere, "energopower" seeks to attend to the contributions of fuel and electricity to the possibility of modern life and its ways of knowing and being. Each of these power concepts is thus more a gestural shorthand than a name for a thing in the world.[25] This makes sense if we supplement our typical English consideration of "power" as noun with the referentiality of the French *pouvoir* or the Spanish *poder*, which in their modal forms indicate the ability to do something—*enablement*. These are forces that allow other forces to happen. Enablement is indeed critical to my perspective on power in this volume. I am interested in what these power concepts enable us to understand about enablement in the world.

With this focus on enablement in mind, I address each of these three categories briefly by turn.

CAPITAL

Kapital for Marx was a dimension of the objectification of human labor power, specifically a result of the manner in which the division of labor severed labor's capacity to channel human will in the development of the self.[26] Instead of an ideal dialectical process of self-realization through productive activity, "capital" signaled how the division of labor allowed labor power to congeal in such a way that it could be alienated from its source, circulate beyond the self, be appropriated and commanded by others, and thus be transformed into new social and material forms. Capital was, in this way, a means of remote enablement (yet one that was always enabled *de infra* by labor power). One thing that capital enabled was the emergence of a class of capitalists who parasitized the labor power of others. But capitalists were not puppeteers—creatures of will and reason—in this paradigm; they were more like mushrooms sprouting on a rotted log of alienated labor power, which was the true enemy that communism sought to oppose. Nevertheless, once capital was set into motion on a mass scale and stabilized by institutions such as money and wage labor, quantifiable, appropriated labor time became the logic of social value in modern society. As Marx wrote in the first chapter of *Das Kapital*, "As values, all commodities are only definite masses of congealed labor-time." In other words, commodities—useful things—were, in effect, masses of capital.

But "congelation," from the Latin verb "*congelare*" (to freeze together), is a slightly misleading translation of the actual noun Marx uses, "*Gallert*," which refers to a gelatinization process in which different animal substances with the potential to yield glue (e.g., meat, bone, connective tissue) are boiled and then cooled to produce a "semisolid, tremulous mass, . . . a concentrated glue solution."[27] Rather than a freezing together of independent parts, "Gallert" suggests an ontological transformation accomplished by adding and then subtracting thermal energy—a recipe of different fleshy forms rendered through heating and cooling into a single sticky material: human labor, in the abstract, binding commodities, people, machines, and "nature" together with its glue. Indeed, this glue potential was unlocked by the thermal rendering process itself.

Paul Burkett and John Bellamy Foster have argued that there was a powerful energo-metabolic substrate to Marx's theories of labor power, alienation, and value extraction. On the topic of surplus value, they write,

> Of course, this value (energy) surplus is not really created out of nothing. Rather, it represents capitalism's appropriation of portions of the *potential* work embodied in labor power recouped from metabolic regeneration largely during non-worktime. And this is only possible insofar as the regeneration of labor power, in both energy and biochemical terms, involves not just consumption of calories from the commodities purchased with the wage, but also fresh air, solar heat, sleep, relaxation, and various domestic activities necessary for the hygiene, feeding, clothing, and housing of the worker. Insofar as capitalism forces the worker to labor beyond necessary labor time, it encroaches on the time required for all these regenerative activities.[28]

Seen in this way, capital becomes an appropriation, quite literally, of fleshy power, a sapping and storage of the regenerative potential of being.

In the *Grundrisse* and the second volume of *Das Kapital*, Marx outlines a more differentiated understanding of capital's forms and also a model of the dialectical development from circulating capital toward fixed capital and of fixed capital toward automated machinery. That machinery constitutes at once infrastructure for the production of use values and also, in the manner of Gallert, a potential energy storage system, gathering and holding productive powers in technological suspension. In Marx's vision, capital strives across its historical development to make itself independent of labor, to be able to absorb the productive powers of labor into itself. As one might expect of the vis viva of bourgeois political economy, capital seeks its liberty. The development of fixed capital—the part of the production process that retains its use form over a period of time rather than being wholly consumed in a production process—is the first stage of this process. Marx emphasizes that durability is crucial: fixed capital must *durably* stand in for direct human labor. The more decisive phase is the movement from fixed capital toward automated machinery, a productive apparatus that operates mechanically according to human design and in which "the human being comes to relate more as watchman and regulator to the production process . . . instead of being its chief actor."[29]

Automation not only advances capital's desire to durably emancipate itself from labor but also precipitates the final paradox between exchange value and use value that Marx believed would necessitate the eventual collapse of the capitalist mode of production.

> On the one side, then, [capital] calls to life all the powers of science and of nature, as of social combination and of social intercourse, in order to make the creation of wealth independent (relatively) of the labor time employed on it. On the other side, it wants to use labor time as the measuring rod for the giant social forces thereby created, and to confine them within the limits required to maintain the already created value as value. Forces of production and social relations—two different sides of the development of the social individual—appear to capital as mere means, and are merely means for it to produce on its limited foundation. In fact, however, they are the material conditions to blow this foundation sky-high.[30]

In other words, infrastructure stores the productive energies of labor in such a way that they can be released later in magnitudes that appear to transcend nominal inputs. Technology, as productive infrastructure, thus appears to be capable of generating and distributing use values with limited need for direct (human) labor power. Once the mass of humanity has those means of production at hand, "the worker," as such, can disappear and with it the alienation of labor and the capital it breathed life into.

In its bridging of labor, productivity, and infrastructure, capital remains a quite generative power concept in the Anthropocene, but perhaps not always in the standard critique-of-capitalism mode. Capital is a critical concept, to be sure, but it must be noted that capital also eventually plays an emancipatory role in Marx's dialectical history, one that seems very much in line with contemporary scientific and political faith in the capacity of technology—and its hidden magnitudes of transformational labor power—to offer us salvation from anthropocenic damnation. Marx could almost be recruited for the accelerationist camp in today's debates.[31] But this also reveals what is problematic about classical Marxist analytics from the perspective of today's damaged planet. Undoubtedly Marx was more attuned to metabolic and energetic questions than is often recognized, but there is no obvious "limit to growth" in his model. Anthropolitical and technopolitical domains seem to have potentially limitless powers at their command. The blind spot of Marx's theory of capital is that he does not account for how machine labor is fueled in the first place and what the ecological consequences of all that fuel use might be. His story of the transformation from human to machine labor does not incorporate adequate attention to what empowers, in a physical sense, the propulsion of use value generation in the machine world. Any utopian project—whether automated cars or renewable energy—that positions technology as the means for enabling the perfection of modernity draws deeply from the conceptual well of capital.

As a final note, I will have little to say about capitalism in this volume. It has a descriptive presence, of course, but I find it to be a red herring analytically. As Kaushik Sunder Rajan has argued, to speak of capitalism in the singular "is an absurdity."[32] I also do no more than gesture toward the concept of the "Capitalocene."[33] I agree here with Dipesh Chakrabarty that there is a necessary division of labor between the analytic work of the "Anthropocene" and the "Capitalocene." Both terms tell important yet partial truths about humanity's history of geological and ecological impacts.

BIOPOWER

"Biopower," in Foucault's original articulation,[34] in the elaborations of his philosophical and sociological interlocutors,[35] and as it appears in the work of anthropological writing on power,[36] signals the consolidation of concepts of life, sexuality, and population as objects and methods of modern governance. In a discussion of the related concept, "governmentality," Foucault shows that life and population became both means and ends of modern political power: "Population comes to appear above all else as the ultimate end of government. In contrast to sovereignty, government has as its purpose not the act of government itself, but the welfare of the population, the improvement of its condition, the increase of its wealth, longevity, health, and so on; and the means the government uses to attain these ends are themselves all, in some sense, immanent to the population."[37] Foucault denies the singularity and separability of the means and ends, causes and effects, bodies and knowledges, instruments and environments, and of course subjects and objects of modern power. But neither is he satisfied with a dialectical portrait of contingency in which subject-object relationality is held in a mutually constitutive dynamic as in the relationship between labor and capital. Foucault's power concepts all denote networks of enablement composed of links and relays that cannot be analytically reduced below the level of a circuitry of forces and signs; in other words, the "apparatus" (*dispositif*) is a "system of relations" between "heterogeneous elements" including corporeality, ethics, discourse, institutions, laws, administrative procedures, and scientific knowledge.[38] This is the operational architecture of *biopouvoir*.

In this respect, the concept of biopower is an extension and refinement of Foucault's general model of modern pouvoir. In *Discipline and Punish*, for example, he contrasts the distributed, discursive, and productive nature of modern power from the more centralized, excessive, and repressive character of sovereign power.[39] Then, in the *History of Sexuality*, Foucault defines biopower as a discursive concentration on sexuality, reproduction, and life: "*Bio-power* . . . designate[s] what brought life and its mechanisms into the realm of explicit calculations and made knowledge-power an agent of the transformation of human life."[40] Rather than a Victorian repression of sexuality, Foucault stresses the relentless signaling, voicing, policing, and measurement of sexual instincts and activities that occurred during the Victorian period as the biopolitical organization of modern governance became increasingly sophisticated and detailed in its operation. He proposes biopower is much the same spirit as this volume does, not as a theory of political ontology but rather as a political analytics capable of mapping a network of modal enablement.

Paul Rabinow and Nikolas Rose argue that Foucault's concept work on biopower was both incomplete and historically specific; that is, it was a way of denoting the gradual conjoining of two force clusters during the eighteenth and nineteenth centuries in Europe. The first force cluster was the anatomo-politics of the human body, "seeking to maximize its forces and integrate it into efficient systems," while the second was "one of regulatory controls, a biopolitics of the population, focusing on the species body, the body imbued with the mechanisms of life: birth, morbidity, mortality, longevity."[41] Rabinow and Rose then offer their own more precise and generalizable formulation for "biopolitics": "The specific strategies and contestations over problematizations of collective human vitality, morbidity and mortality; over the forms of knowledge, regimes of authority, and practices of intervention that are desirable, legitimate and efficacious." One notes immediately that this formulation, like Foucault's original, is anthropolitical, and in Povinelli's terms, "biontological." Biopower and biopolitics specifically concern the management and control of a human vitality that is distinguished doubly from a domain of nonhuman life and nonlife more generally. Rabinow and Rose, and many other anthropologists besides, have effectively retooled the biopower concept for twentieth- and twenty-first-century conditions by bringing together the sciences, politics, and economies of life, where "life" itself involves issues as far ranging as sexuality, reproduction, genomics, infrastructure, population, care of the self, and indeed "environment."[42]

Still, life in the Foucauldian analytical imagination clearly centers on human life. This close anchorage to "the human," even as it denies the authoritative overtures of "humanism," is, I strongly suspect, one reason that the concept has proven so compelling among anthropologists as a way of gaining traction on political power. Another reason is that the analytics of biopouvoir, especially if generalized as Rabinow and Rose have done, are remarkably flexible and adaptable to almost any circumstance of governance. More than this, biopower captures rather elegantly many salient features of political power today, especially interventions of expertise and authority concerning health, security, and population.

In Mexico, as we will find in the ethnography, governmental discourse on renewable energy development is deeply saturated by biopolitical reasoning in two respects. First, there is the abundant *environmentality* expressed by the federal government and renewable energy developers in the project of climate change mitigation, a project that is also tied to securing the safety of Mexico's population as new vulnerabilities to drought and flooding are exposed.[43] Second, there is the administrative concern to use wind resources to stimulate the circulatory economy of the Isthmus of Tehuantepec, to attract new resources for development, and to provide new opportunities and infrastructures for commerce, education, and health. Especially in the discourse of state and local governance, the project of wind power development is consistently articulated in biopolitical terms: as means of guaranteeing or improving the health and welfare of human environments, economies, communities, and individuals.

Yet the phenomena of the Anthropocene challenge the anthropolitics and technopolitics of the biopower concept as well. The shockwaves set off by overuse of carbon and nuclear energy—from rising seas to environmental toxicities and nuclear tragedies—have shaken the foundations of contemporary biopolitical regimes in such a way that we find fissures opening and fuel (sometimes quite literally) flowing into the groundwater of bios. This is why I believe the time has come to add "energopower" to our roster of concepts for analyzing political power. Foucault, I think, would approve in that his genealogical method was not designed to inquire into timeless conditions that endure throughout history but rather to examine "the constitution of the subject across history."[44] That is to say, if "biopower" is one of our most enabling keywords for analyzing political power today, it also seems appropriate in the spirit of Foucault's original intervention to subvert it through new genealogical exercises lest we come to convince ourselves that "biopower" denotes some transhistorical form of modern power and subjectivity.

ENERGOPOWER

As a power concept, energopower draws attention toward the impacts of fuel and electricity upon the domain of the anthropolitical, including biopower, capital, and all its other force clusters. This line of thinking is by no means entirely new. I have argued elsewhere that anthropology and the human sciences have been punctuated by periods of very generative thinking about energy, particularly around times when there were widespread perceptions of energy transformation and/or crisis.[45] In the 1940s, for example, the coming into being of atomic energy precipitated both cornucopian and dystopian thinking about new energic plenitudes, their luxuries and dangers.[46] In the 1970s, the reorganization of the geopolitics of oil and the experiential crisis of the "oil shocks" helped stimulate another period of thinking about energy, but one that was largely eclipsed again in the 1980s as Reaganism, neoliberalism, and finance capitalism stole center stage with promises of a return to prosperity and security.[47] Still, the technocratic modernization narratives of the 1950s and 1960s were irreversibly disrupted in the 1970s, and feminist and post-Structuralist critiques of technoscience were the first tremors of the broader antianthropocentric turn that the human sciences are experiencing today.[48] As global warming, climate change, and other anthropocenic phenomena became more actively mediated and more epistemically present in the first decade of the twenty-first century, energy has once again begun to spark at the margins of social theory.[49] Two theorists in particular have helped to give shape and content to the ethics and epistemics of what I am terming "energopolitical" analysis.

The first is energy futurist Hermann Scheer and his call for a decentralized "solar economy."[50] Scheer was one of the chief architects of Germany's Energiewende (renewable energy transition) and cowrote Germany's much-imitated feed-in tariff law that forced German utilities to guarantee long-term purchase agreements for renewable energy to create stability and incentives for solar and wind power producers. The effect of this policy intervention was an unexpectedly rapid shift toward renewable energy production in Germany. In large part thanks to the stimulus initially provided by Scheer's feed-in tariff legislation, in 2017, of the 654.1 terawatt-hours of electricity in the German national grid, 16.1 percent came from renewable resources and 15.1 percent from wind power alone.

However, there was immense political resistance to Scheer's plan in the beginning, and his analysis of the various obstacles that rapid renewable energy transition faced helps to surface how energy infrastructure—particularly fuel supply chains and electricity transmission systems—exert a massive, hidden influence over political and economic systems. Scheer pointed to the adaptation of global and national economies to the "long supply chain" infrastructures characteristic of fossil and nuclear fuel resources. Scheer viewed twentieth- and twenty-first-century globalization as largely driven by the extraction and control of these fuels. He observed that long energy supply chains are, in their material nature, inefficient and thus demand allied infrastructures of translocal domination to guarantee unimpeded flows of critical resources. This domination imperative has deeply informed geopolitics even when it is masked by nationalist discourses of security and well-being, by post/neo/colonial missions of civilization and development, and most recently by the utopian logic of a self-regulating market.

Solar energy—whether in its direct form of insolation or in the indirect forms of wind and biomass—has the physical advantages, Scheer argues, of ubiquity and superabundance, thus allowing for more efficient and decentralized short supply chains that are also more susceptible to democratic political control: "Shorter renewable energy supply chains will make it impossible to dominate entire economies. Renewable energy will liberate society from fossil fuel dependency."[51] Recognizing that reliance upon fossil and nuclear fuels has driven the world toward anthropocenic ruin, Scheer challenges the assumption that it is important to maintain large-scale power grids and pipeline systems at all. He calculates that even energy-intensive modernity can be maintained purely on the basis of small-scale solar, wind, and biofuel resources given contemporary technologies. The resistance to infrastructural transformation thus has less to do with the fear of blackouts or "energy poverty"—although societal paralysis and devolution continue to be conjured to delegtimate renewable energy transition—but rather because of a more basic but also invisible codependence between our contemporary infrastructures of political power and our infrastructures of energy. This is to say that translocal high-voltage grids and fossil fuel infrastructures—both products of early twentieth-century political and industrial concentration that was enabled, in turn, by the burning of fossil fuels—evolved over the course of the twentieth century. They became primary instruments for the monopolization of political authority, thus constituting what I term an energopolitical apparatus reinforcing both the inertia of a particular organization of fuel and a particular organization of state-based political power. This convergence generated an energo-material path dependency, according to Scheer, one that resists the imagination of alternatives to the long-chained fossil-fueled status quo. For to imagine an alternative to "the grid" is, in essence, to imagine an alternative to centralized political authority, bureaucracy, and "the state" as well.

Scheer's analysis shows how a power concept like energopower has the capacity not only for critical traction on past and contemporary entanglements of fuel, electricity, and political power, it also has the capacity to provoke discussion as to how emergent energic infrastructures could contribute to the development of new forms of modern political and social experience. Scheer's insistence on locally sourced, owned, and managed electricity echoes today in a surge of community-owned renewable energy projects worldwide. And, as we see in chapter 1, such initiatives belong to Mexico's aeolian politics as a community-owned wind park in Ixtepec struggles to come into being against the energopolitical apparatus and inertia of the electricity parastatal, CFE (Comisión Federal de Electricidad), as well as the interests of transnational green capitalism, which has claimed the lucrative Oaxacan wind market for itself. In all such instances, "energopower" gives us a way to join together discussion of emergent postneoliberal political potentialities with the energic forms of "revolutionary infrastructure" that will necessarily enable them.[52]

The other great inspiration for "energopower" comes from Timothy Mitchell's prescient and influential Carbon Democracy project.[53] No stranger to biopolitical analysis, Mitchell digs deeply into the history of carbon energy to surface the dependency of modern democratic power upon carbon energy systems: first coal, later oil, and now natural gas. Much like Scheer,

Mitchell begins with the contemporary dependence of modern Northern life upon massive energy expenditure. He then retraces the way that the harnessing and organization of fossil fuels has shaped the trajectory and forms of modern political power. Mitchell shows, for example, how the consolidation of social democracy in the late nineteenth century crucially depended on the materialities and infrastructures of coal that allowed miners to establish chokepoints in fuel flows, which exerted immense pressure on dominant political and capitalist institutions until they eventually acceded to labor reforms.[54] He then links the biopolitical norms of twentieth-century Keynesian welfarism to a regime of expertise that characterized oil as an inexhaustible and increasingly inexpensive resource that was capable of fueling the endless growth of national economies. This was certainly true in Mexico, where Lázaro Cárdenas's nationalization of Mexico's fossil fuel resources helped propel major biopolitical investments in the midtwentieth century. Mexico also benefitted from the oil shocks of the 1970s as the country became a key partner in the global North's effort to reestablish secure flows of fossil fuels. Still, in Mitchell's argument, "growth" and "economy" ultimately reveal themselves to be tokens of petroknowledge, whose apparent truthfulness was owed to a midtwentieth-century geopolitics of neoimperial control over the Middle East and its subsoil resources.[55] When that control ruptured with the formation of OPEC, the foundation of Keynesian biopolitical authority disappeared rapidly. Growth declined radically across the global North, and a different configuration of life and capital—the one normally glossed as "neoliberalism"—exploited the crisis to assert its dominance. Although those politics consistently vowed a resurrection of Keynesian growth patterns, the historical record shows those promises to have been deliberate or accidental lies.[56]

In keeping with Mitchell's and Scheer's analyses, the renaissance of finance capitalism after the 1970s can be viewed as an effort to maintain value flows through the channels grooved by Anglo-American petrohegemony (not mention, of course, earlier colonial and imperial relations), an "oil standard" replacing the "gold standard."[57] But rentier financialism, unsurprisingly, lacked Keynesianism's investment in biopolitical development. Finance was, in the end, a more obviously parasitical method of extracting and consolidating value and one that perhaps could be judged weaker in terms of the energies it commanded directly. Keynesianism, as a state-centered political order, had the machinics and materialities of industrial petropower at its disposal, which allowed for massive projects of infrastructural and capital development, regardless of what purposes those projects were meant to serve. Finance capitalism, on the other hand, could only indirectly access the centralized authority of the state and the powers of industry. In its core practice, it had to make do with the electric speed of information transfer and the opportunities for arbitrage that those created. Of course, as finance more securely positioned itself as the central nervous system of globalization, it was able to reorient and minimize biopolitical priorities through the leverage of debt and the constant threat of capital withdrawal.[58] Although finance capital is by no means intrinsically hostile to infrastructure—the internet is an excellent example—its relationship to public infrastructure is at best ambivalent and more often directly or indirectly critical. We have seen that tension revealed of late in a wave of integralist infrastructural nostalgia for a period before finance capital became ascendant.

The neoliberal disarticulation of the energopolitical capacities of the global North was masked by the incremental pace of the dissolution of public infrastructures, by the popular utopias of internet and real estate bubbles, and by fear- and warmongering designed to maintain attention elsewhere.[59] But this masquerade existed only for the global North. In countries like Mexico, neoliberal policy regimes and structural adjustment policies led swiftly and obviously to misery for entire nations, especially for those strata that had gained prosperity through the exercise of Keynesian biopolitics. The financial crisis of 2007–9 gave the North a taste of what the South had been experiencing for three decades already. It left the dominant ideological order discredited even though it left the actual institutional apparatus of finance relatively untouched. Now, almost a decade on, there is a feeling of living in ruins, both infrastructural and imaginational.[60]

Perhaps those ruins will prove to be the fertilizer of something else, whether the return to fossil-fueled national glory dreamed of by Trumpists and Brexiteers or the solar emancipation aspired to by Scheerians the world over. Whatever intermediary forms of postneoliberal life we are now witnessing, the eschatology of the Anthropocene suggests that the further pursuit of growth as prosperity—whether Keynesian, neoliberal, or otherwise—points only deathward. This double bind remains powerfully suppressed since even our ready-made idioms of revolution tend to depend on massive energic magnitudes.[61] The North has not yet found a way to imagine low-energy prosperity, freedom, and happiness. My argument is simply that energopolitical analysis offers a different set of analytical attentions than those of biopower and capital and, as such, may help enable us to tell different stories and imagine different futures.

It is in this spirit that this volume is titled *Energopolitics.* I do not mean to suggest that energopower is the most important of our conceptual minima. Rather, I put forward energopolitics as a general project of inquiry, as a hashtag if you will, for a conversation that I believe would be worthwhile to pursue in greater depth. If Mitchell's project offers a deep and rich political history of a specific trajectory of energic materiality and infrastructure conditioning anthropolitical emergence, then "energopower" offers a more general power concept that can serve to bring into juxtaposition many such cases of enablement, including the study of wind, land, and power offered in this volume. It is hopefully clear from my discussions of capital and biopower that rethinking them as power concepts focused on enablement helps to unlock their own energopolitical storylines as well. Finally, terminologically speaking, "energopolitics" joins together the modality of "pouvoir" with the Aristotleian ἐνέργεια (*enérgeia,* "activity"),[62] which was later redefined by modern physics as "work," most often as the capacitation for (mechanical) work. Although I find a narrow definition of "energy" as "work" conceptually disabling in many respects, combining "energy" in its capacitational sense with "pouvoir" creates a kind of double modalization in the term "energopolitics" that helpfully gestures toward the multiple and nested modes of enablement that I am seeking to map in the ethnography.

MORE MINIMA: DESIRE, ONTOPOWER, . . .

Such multiplicity suggests the need to broaden the set of conceptual minima beyond a conventional triad. In brief: please do. This set should remain open for addition and exploration. Two additional concepts that I have found valuable in this analysis are the psychoanalytic (Freudian) concept of "desire" and Brian Massumi's Deleuze-inspired "ontopower."

I find the ontopower concept valuable less in terms of the militarization process that Massumi has recently documented at length and more in terms of the concept's gesture toward a force cluster of affective "living powers" that exceed but also inform human political power.[63] The winds of the Isthmus of Tehuantepec, for example, are ontopolitical, as anyone knows who has turned a corner and been knocked to the ground or has seen tractor trailers flipped over on the highway between La Ventosa and La Venta. Those winds are not reducible to any project of human political imagination or organization, although human beings have long sought to capture their powers, whether with words and songs or, more recently, with blades and turbines. Across the world, aeolian ontopowers have come to have a special allure for anthropolitical projects of renewable energy development. But, at best, those projects seek to harness a force they know cannot be controlled or administered fully.

In the Isthmus, *binnizá* (Zapotec) people have historically equated the winds with the cosmological force of life itself, *bi*, and when Christianity came, the north wind offered enough of a challenge to the power of the divine that it became known as "the devil's wind." Today, istmeños respect the power of *el viento viejo* (the old wind) at least as much as they respect local political power and certainly more than the translocal political power of the Mexican state. El viento is not a power that hides away, seeking to exert influence from afar; it is experienced more as a medium that courses around locals at all times, by turns irritating with gust-borne gravel and offering relief from subtropical swelter.

Freudian desire (*Wunsch*), meanwhile, may be less obvious as a power concept, but in the context of thinking about pouvoir/enablement, desire offers a very valuable and specific insight. As I have written at greater length elsewhere,[64] Freud's late neurology and early metapsychology were strongly influenced by thermodynamics and electrical research, articulating a model of psychic operation as a largely homeostatic energy system managing exogenous and endogenous stimuli to maintain a tolerable load of excitation. The relationship between primary process and secondary process is the crucial dynamic. The primary process represents the psychic apparatus's effort to reduce excitation that has been created by unconscious charging of memories into hallucinatory identifications. The psychic apparatus strains, irrationally to its core, to repeat past acts of need satisfaction, reducing pains of want through the pleasures of imaginary discharge. In other words, one searches always for that excessive pleasure of infantile satisfaction, drawing available objects and subjects into one's field of desire even when it is not clear that they can offer any fulfillment whatsoever. To reduce this innate hallucinatory tendency, the primary process is interrupted by a secondary process of social-environmental conditioning that seeks to channel the search for pleasure instead through the intricacies of language and custom. The fact that the secondary process must continuously seek to repress and deflect the primary process creates a fundamentally entropic condition in the psychic apparatus. In instances of psychosis, neurosis, and dreaming, Freud believes we see how the weakening of secondary defense mechanisms allows the energy flows of the primary process to more directly influence the systems of consciousness and perception in the form of hallucinatory imagination.

The concept of desire contains this drama within itself. On the one hand there is a mad pursuit of pleasure and fulfillment, on the other a constant attempt to temper and deflect the urgency of pursuit toward thoughts and behaviors that can be reconciled with our fundamental sociality in the form of social norms and, indeed, reason. But the primary process remains primary. It is the volcanic power of hallucinatory identification—unconscious belief that the pursuit of one's present objects of desire will result in repetition of the excessive pleasure of past satisfaction—that propels us forward. Desire remains perpetually unfulfilled because of its past orientation: it rejects the possibility of unknown forms of future satisfaction and pleasure. Instead, it constantly tries to commensurate and manipulate contemporary encounters to suit its (again, mostly unconscious) memory archive. It is thus possible to view desire as endless and eternal and, as Žižek writes, to see its ultimate function as self-reproduction: "Desire's *raison d'etre* . . . is not to realize its goal, to find full satisfaction, but to reproduce itself as desire."[65]

In this respect, I find the concept of desire invaluable in accounting for the apparent paradox that the global North continues to utilize ecologically toxic magnitudes of fossil fuels despite some level of rational awareness that this is not a good thing. This is precisely the paradox of desire: a backward-looking investment in past pleasure always seems to trump consciousness and reason precisely because the future has no memories to offer. Such desire continues to enable itself in pursuit of the past pleasures of carbon modernity. Likewise, when governmental actors and renewable energy activists, including the Hermann Scheers of the world, promise that a clean energy transition can be accomplished without loss and without sacrifice, one sees there, too, an attempt to define the future in terms of a memory archive constituted by the energic abundance of petropower. Even as we work to shift rationality toward a critical and transformational engagement with the Anthropocene, the concept of desire teaches us to respect the primary process that will do anything in its power to pull the future into the gravitational orbit of the past. It is a humbling reminder of the limits of reason to steer us toward a future that does not repeat the past. Somehow, we also have to create memories of the future that we hope to attain.[66] This is why, to my mind, the work of the arts is so vital to unmaking the Anthropocene.[67]

Ethnographic Maxima

We these minima laid out, let us move from concepts to ethnography. As outlined in our joint preface, Cymene and I believe Mexico to be one of the richest and most rewarding cases of the ecological, social, and political complexities of renewable energy transition across the world today. My ethnographic strategy in this volume is to structure the presentation of our fieldwork in such a way as to locate those situations, encounters, and relations where the conceptual minima are absolutely necessary to understand what unfolded in the course of our research. But I also spend a great deal of time exploring the abundant force clusters within southern Mexico that cannot be derived from concepts like capital, biopower, and energopower. We have termed the full ecology of these clusters surrounding wind power development "aeolian politics" as a way of defamiliarizing wind power as a more conventional object of contestation or salvation.[68]

Southern Mexico's aeolian politics include, for example, a complex and contested history of land tenure in the region around Juchitán, whose legacies exert a constant influence over wind power development. There is the brokerage and lobbying work of NGOs and fixers and unions. There are project developers and financiers constantly laying groundwork for securing generous financial returns on their investments. There are the clientelist networks and corporatist machinations of the Mexican political parties, in particular the PRI (the Institutional Revolutionary Party) and PRD (the Party of the Democratic Revolution), which often seem much more vital than the governmental bureaucracy they inhabit. There are the logics of caciquismo (boss politics) and student/teacher/peasant/worker/fisher opposition movements operating in the isthmus, whose principles both inform and exceed the political parties. There are historical tensions between Mexico City and Oaxaca City, between Oaxaca City and the Isthmus of Tehuantepec, and between the istmeño towns and the surrounding countryside, all of which must be taken into account. There are historical rivalries between the binnizá (Zapotec) and ikojts (Huave) peoples whose lands, seas, and winds are all affected by wind parks. There is a federal government that is anxious about waning petropower and climate change, a state government that is anxious to perform its own sovereignty, and a vulnerable parastatal electricity utility (CFE) that is trying to stave off privatization. There is the infrastructural inertia of an electrical grid system that has been optimized for fossil-fueled thermoelectric energy supply. There are the electrical engineers and grid administrators whose expertise is likewise optimized to manage baseload thermoelectric supply and who are skeptical of wind power for its intermittency. There are legacies of settler colonialism and racism that shape both the logic of governmental intervention as well as conceptions of local and indigenous sovereignty. There are the promises and entitlements of the Mexican Revolution, which have not been forgotten. There are alliances and disjunctures between federal and state bureaucracies in Mexico. And not least, there is the power of the wind itself, which howls and "jams corncobs in your nose" as the binnizá poet Victor Terán writes. In their cross-drafts and swirls, these many forces provide aeolian politics with its turbulent vortex.

This is also no closed set, but all these forces belong to the maxima of an adequate anthropological analysis of wind power development in southern Mexico, and they are given their due in the chapters that follow. The importance of local, regional, and national scales of enablement vividly reminds us that anthropology of political power in the Anthropocene will always need to look beyond conceptual minima to comprehend its scenes of engagement. Fieldwork speaks *terroir* to *pouvoir*, highlighting the modal multiplicity of enablement that inheres in any situation of human life and endeavor.[69]

For the same reason, I have organized the ethnographic narrative as a journey between key places that we visited in the course of fieldwork to show how the political terroir varies substantially from site to site, destabilizing glosses like "political culture" or "wind power development" or "southern Mexico" while reinjecting difference and locality into them. Often in what follows, we shall see that the politics of land in the isthmus play an especially salient role in constituting local terroir. The isthmus contains at least three distinct land tenure regimes: the *bienes comunales* of ancestral indigenous communities, the *bienes ejidales* (ejidos) granted to landless peasants after the Mexican Revolution, and a heterogeneous array of forms of private land ownership—sometime de jure and more often de facto. None of these legal regimes were designed with the facilitation of energy "megaprojects" in mind. Projects of wind power development that seek to shepherd the isthmus from an agrarian past to a postindustrial future have thus had to navigate this uneasy terrain, seeking to satisfy communal and private owners without incurring the animosity of neighbors and inciting factionalism.[70]

Our passage begins in Ixtepec (chapter 1), where an NGO and a group of *comuneros* (indigenous communal landholders) sought to create Latin America's first community-owned wind park, Yansa-Ixtepec, on communal agrarian land. Had it succeeded, this project would have ruptured a cozy arrangement of transnational capital, biopolitical aspiration, and energopolitical infrastructure that has put Oaxacan wind power into overdrive in the past several years. The chapter surfaces the alternate political and social imagination wedded to Yansa-Ixtepec and the way it gave voice not only to different biopolitical expectations but also to a desire to reconnect to land and to strengthen indigenous sovereignty. We also explore at length the actors and infrastructures that strenuously opposed Yansa-Ixtepec's effort to come into being.

We then move east to La Ventosa (chapter 2), a town nearly wholly encircled by wind parks that have been constructed in the dominant private-public partnership (PPP) model. We examine the logic and history of that model in detail but also the reasons that La Ventosa became its epicenter. Caciques (bosses) control the town, it is said, abetted by influential party-political networks and relationships with transnational energy companies; it was the bosses who decided that wind power was La Ventosa's future. Dreams of broader white-collar prosperity for the town compete with machinations to expand political influence and to secure substantial rentier incomes. Walking the streets of La Ventosa for a house-by-house survey of community opinion, we came to understand the deep ambivalence that most La Ventosans feel about the rapid transformation of their lived environment, this "development" that will shape their community for decades to come.

The Pan-American Highway takes us upland from the Isthmus of Tehuantepec to Oaxaca City (chapter 3), where we find a state government in disarray concerning wind power. The Oaxacan state finds itself cut out of the developmental loop by alliances between federal government agencies, transnational developers, and istmeño political leaders. Yet these same forces blame the Oaxacans for their failure to manage the rising tensions and violence surrounding the wind parks. As the government struggled to govern, we witnessed its agents participating in a variety of forms of "performative sovereignty" designed to project more secure control over the future of wind power development. However, the growing recourse to characterizing the isthmus as an eternal indigenous other, beyond the mestizo state and nation, could also be taken as a frank admission that it might be beyond the capacity of the Oaxacan Valley to influence, let alone control, what happens in its historically renegade province.

Heading farther northwest, we eventually come to the locus of national political power, Mexico City (chapter 4), where we met the agents of bio-power, capital, and energopower most closely affiliated with designing and enabling wind power development in the isthmus. We met a caravan of activists, a seasoned journalist on the energy beat, several engineers from the parastatal electricity utility, the deputy minister of electricity, the director of the federal energy regulatory agency, a transnational banker with millions invested in istmeño wind parks, and a former Oaxacan governor and party kingmaker who has become a fixer for the wind industry. We came to realize the extent to which (a) these agents and their agencies often work at cross-purposes to one another—thus shattering the often-mediated image of a consolidated federal political policy regarding wind power development—and (b) how little most of them know about the isthmus, its residents, and their reasons for supporting and opposing the parks.

Finally, we return again to the isthmus, our passage ending in the center of aeolian politics, Juchitán (chapter 5), where alternative futures of istmeño wind power burn brightly in conflict with one another. We discuss how contemporary aeolian politics have been informed by a long history of Juchiteco resistance to foreign powers and how the history of land tenure and class conflict in the isthmus overshadows thinking about wind mega-projects today. We met the leaders of the local political factions and social movements that organized themselves both for and against the parks. And we discovered how the intrigues surrounding wind and power filtered down from elite machinations into the barrios. One way or another, Juchitecos believe they will be the ones to decisively determine the trajectory of istmeño wind power going forward.

And now, with thanks for your readerly patience, theory stands aside and invites ethnography to do its work.

1. Ixtepec

August 18, 2012. The view as we ride into the hills north of Ixtepec is stunning. As we clear the ranchland outside of the city, moving through the ubiquitous smoldering garbage dumps whose acrid white smoke marks the margins of urban life in the isthmus, a densely verdant landscape folds in upon the road. Sergio and I have to duck frequently to avoid being swept out of the bed of Juan's truck by low-lying branches. When we do not have to share our passage with packs of Brahman cows shouldering nervously against us, Juan drives quickly, and we dig our fingers into the worn wooden railings to hold on. In the hills, brilliant sunlight scatters through the trees' leaves; the air is lush. We begin passing through rivulets that transform the road into a sparkling mirage at a distance. When we hit them at speed, both water and color spray into the air as hundreds of tropical butterflies take flight.

Sergio is drinking it all in. "I always feel so happy when I come up here."

With the wind rushing around us, it is a marvelously *Titanic* moment. And in truth, it is the happiest I have ever seen him. For all is not well with his bold project to orchestrate the first community-owned wind farm in Latin America.

Sergio confides loudly over the noise of the truck and the wind that he is afraid that the Mexican national electricity utility, the Comisión Federal de Electridad (CFE), will not even allow them to bid for grid access because the current tender model requires massive bank deposits and letters of credit. "This is a poor indigenous community. They don't have millions of dollars lying around to stake as collateral," he explains. And then, even if they are allowed to bid, Sergio worries that they may not win the competition because although Sergio and his collaborators in the Ixtepecan bienes comunales (communal estates) have "ambitious social development goals" in mind, CFE is required by federal law to purchase energy at the lowest unit cost.

Social development is the theme of our trip into the hills this afternoon. We are visiting the Ixtepecan township (*agencia municipal*) of El Zapote (population 809) to help cement local support for the project. Sergio explains that El Zapote borders on the footprint of the planned park and that some Zapotero farmers use land that the project needs to lease.

Juan tells us later that the Zapoteros' true interest in the wind park had less to do with land rents, let alone clean energy, than with getting running water for the village, making electricity more constant and reliable, and developing better transport linkages since the villagers had few vehicles of their own.

Indeed, as we near the village, the roads metamorphose from asphalt to gravel to dirt, and we pass many men and some women walking the other way, likely headed the several kilometers into town on foot. The promise of infrastructure is, in other words, said to be the Zapoteros' true desire. Or perhaps more accurately, in El Zapote—as in many parts of the isthmus, long ignored by governmental programs of biopolitical development, often circumvented by flows of transnational capital—there is no way to clearly separate aspirations for improved infrastructures of life from those of energy. The Ixtepecan plan to fuse community-owned clean energy and social development presents these entangled dreams in a particularly vivid way.

"This Isn't Denmark"

Sergio is handsome, wiry, intense, and projects tireless optimism. He was born in Spain and speaks Spanish with a peninsular accent as well as perfect English. After studying in Germany and the UK and spending several years doing community organization work in South Asia and Latin America, in 2006 he ended up working for the Nordic Folkecenter for Renewable Energy in Ydby, Denmark, part of the World Wind Energy Institute network, where he became increasingly interested in how renewable energy could be harnessed as a tool for community development.

It was also a boom time for the wind industry globally, Sergio remembers. "Before the 2008 crash, there was so much more demand than supply that it was difficult for smaller players like communities to get access to the technology they needed." So he founded the Yansa Group with the ambition to export the Danish model of "community wind" production to rural communities in developing countries in order to help democratize access to renewable energy expertise and technology and to serve as a powerful tool for community integration and development.[1]

While in Denmark, he heard about the "wind rush" in Oaxaca through members of his family (his mother is Mexican), but he also learned of the rising resistance and violence surrounding wind park projects. He travelled to Oaxaca briefly in 2008, spoke with residents of several communities likely to be impacted by wind parks, and then travelled to Mexico City to meet with the president of the industrial lobbying organization AMDEE,[2] which was spearheading wind development in Oaxaca.

"Nowadays they've got a more polished message, but back then, what I was hearing from them was really outrageous, blunt, even racist. They viewed the communities as villains, ignorant people ruled by small leaders who wanted bribes and were stopping progress. I told him that in other parts of the world, like Denmark, communities were being engaged more constructively as partners in wind development."

The president scowled at him, and the meeting soured to the point that, according to Sergio, it ended with a thinly veiled threat: "I don't know what you're going to do with all this information, but I'd be careful about what you wrote. This isn't Denmark. Anyone can fall off his horse here."

But Sergio was in no way dissuaded. He relishes confrontation with the forces of industry and government, which he sees to be deforming and corrupting the course of wind development in Mexico. So he started traveling frequently from his apartment in Mexico City to the isthmus, connecting with some of the activists working against the wind parks, who in turn had networks in the communities that were being impacted.

"One of the first things was to try to shift their perspective on wind energy. For a lot of people it had become something evil; it meant giving your land to Spaniards. I told them that this wasn't the way it was in some other countries. There could be community-owned wind energy, even in Mexico. At first, they thought I was out of my mind. But I kept repeating it and repeating it."

By chance, then, one of the activists happened to share a bus ride with a comunero from Ixtepec, who told an interesting story about how their comuna was trying to convince CFE to let them build a community wind park.[3] The electric company needed to use Ixtepecan land in order to build a new substation to evacuate all the electricity being generated by wind parks to the high-voltage arteries of the national grid. Spending an estimated $116 million on the project, CFE was making the largest federal infrastructural investment in the region since the 1960s. When CFE presented the comuna with its substation plans, some voiced the idea that Ixtepec should get a community wind park in exchange to raise revenue for the city.

Sergio explained, "There were a couple of visionaries there, and they were able to at least convince CFE to pay for them to meet with consultants on community development. After those meetings, the comuna representatives drafted a plan for a community wind park and tried to get CFE to agree to it."[4]

But CFE's response was, "You guys are out of your mind, you'll never find the money for that, so we won't reserve space on the grid for you."

"And CFE didn't." Sergio went on. "But the idea didn't go away, and when my activist friend told them that there was this crazy Spaniard running around saying that community wind parks were possible, it clicked, and we met, and have been working together since 2009."

Sergio positively contrasted the political location for community development in Ixtepec with that of the district capital, Juchitán, the epicenter of the wind rush. "Much of the wind development in the isthmus is taking place on land that traditionally belonged to Juchitán's bienes comunales. But with the rise of COCEI and the political violence of the 1970s and 1980s,[5] the comuna leadership was broken and hasn't reformed for decades. But the land is still legally communal. So, in a sense, all these contracts that are being written to lease land around Juchitán for wind parks are completely invalid. But in Ixtepec you have an unbroken comuna tradition."

"And the other thing that's interesting about them," he says, "is that in the 1960s some of the students who helped the farmers stave off land occupations actually were admitted into the comuna in thanks. Usually comuneros in the isthmus are from families who have been farming their plots for generations. In Ixtepec you have those comuneros but then also a leftist intelligentsia, lawyers and engineers, and they are the ones really pushing for a wind park."

Other observers and participants drawn together by the power of the istmeño wind like to credit Sergio as the hero, or antihero, behind the Ixtepecan wind park proposal, but he consistently defers credit from himself and Yansa to the comuna. Sergio was, in his own words, simply "a catalyst." Of course, it is quite possible to view Yansa as not only enabling the realization of spontaneously generated community interests and agendas but also as ideologically shaping those interests and agendas through a particular expertise of "development."[6] Such expertise is further inflected in this case by belief in the capacity of green energy to produce new political and social relations. There is no doubt, as a wealth of anthropological research on NGOs elsewhere has demonstrated,[7] that NGOs exert powerful "performativity,"[8] changing conditions and relations through their expert intervention. Yansa indeed cultivated an energo-bio-power/knowledge that absorbed local interests and knowledges and decisively honed them toward a particular infrastructural realization and social future. And as we shall see later on in this chapter, there were many Ixtepecans who rejected the idea that the agrarian comuna represented their "community" at all. But the general criticism that NGOs tend to impose political ideologies of their own in development programs, while well taken, also does not scratch the surface of the subtle flux of local power relations—including, notably, Sergio's close friendship with his allies in the Ixtepec comuna, those allies' slightly awkward relationship to the "normal comuneros," political factionalism within the asamblea, and the asamblea's alienation from the broader population of Ixtepec—relations that also inflected, accentuated, and negated, by turns, the Yansan imagination.

And for our part, befriending Sergio was certainly catalytic for our field research since he helped us massively in terms of making contacts across the isthmus, especially among those opposed to the status quo of Oaxacan wind development. He also became a partner in many of our more interesting adventures through the windy landscape of southern Mexico. All of which helped us to appreciate how Sergio's impact in the contentious political field of wind power development in the isthmus has been more extensive (and elusive) than many will ever know.

The Yansa Model

Having arrived in El Zapote, the asamblea gathers on the veranda of what appears to be the only public building in town, aptly sheltered by the abundant shade of an old white sapote tree. Its hard green ovoid fruits litter the ground between us as we set up plastic chairs in a rough circle. About twenty Zapoteros gather, older men wearing broad campesino hats, older women in vividly embroidered Zapotec dresses, younger men and women in jeans and T-shirts. Our daughter, Brijzha, investigates the nearby unpaved road while the other children explain to her which ants she can poke with a stick and which are the "snake killers" to stay away from.

The agente municipal introduces himself softly and then turns the floor over to Sergio for his presentation. Although the main agenda for this meeting concerns *el agua* (water) and *transporte* (transportation) and how to organize the labor and funding to get them running in the township,[9] Sergio begins by reading the minutes from a previous meeting focused on "making sure that women are participating fully in the process of establishing a community wind park and that their concerns are added to the social goals of the project." And the last third of the meeting is a devoted to a more detailed presentation of Yansa's proposals for *fines sociales* (social aims) based upon their community *consulta* (a consultation involving 148 interviews in 2012).

Among other questions, interviewees were asked, "If you had 100,000 pesos to invest and had to choose between the areas of culture, education, jobs, environment, recreation, and health, how would you distribute it?" The results averaged out to 11 percent culture, 23 percent education, 26 percent jobs, 12 percent environment, 9 percent recreation, and 19 percent health.[10] Following these priorities, Yansa is suggesting an emphasis on increased opportunities for work (by paving roads, creating public transportation links, and putting in telephone lines), education (by creating a primary school and offering grants for older children to study elsewhere), health (through a clinic but also a recreation area), and finally for *desarollo comunitario* (community development) "to promote a harmonious, sustainable life in El Zapote taking into account both the necessity of ensuring human rights for all and sustaining the natural environment." These points are acknowledged by the attendees' nodding. They also closely correspond with Yansa's published vision of their social development targets for the community wind park.[11]

The resources for social development would come from a unique partnership model that Yansa had designed to connect the NGO, the comuna, development banks, and social investors. The basic elements of the partnership were organized as follows: assuming that Yansa won a CFE tender for grid access, they would immediately form a community interest company (Yansa-Ixtepec CIC), which would own the wind park and negotiate a land lease agreement with the comuna (which would continue to "own" the land upon which the wind park was built).[12] The estimated cost of building the park, consisting of thirty-four 3 megawatt turbines (production capacity of 102 megawatts total), is $200 million. Construction funds would be raised through a mix of 70–80 percent development bank funding and 20–30 percent social investment funds (which would be willing to accept a below-market return on their investments to achieve specific social targets. In a moment of gloomy bravado Sergio once assured us that he could raise those funds "in a heartbeat if we had a contract with CFE. There are so many investors out there waiting for opportunities like this one"). As many construction jobs as possible would be given to comuna members. Once the park was operational, per Mexican law, Yansa-Ixtepec CIC would sell its electricity to CFE for the next twenty years. Sergio estimated the total annual surplus from the park (after servicing debts and interest payments to banks and investors) as around 50 million pesos per year ($3.81 million).

That surplus would be divided fifty-fifty between Yansa and the comuna. Yansa would use its half as seed funding for further community wind park projects elsewhere in the world. As Sergio explained, "The social investors don't want to fund a one-off. They would rather invest in a project that could grow and have a broader impact." The comuna's half would be further divided into approximately 3 million pesos in payments to the campesino *posesionarios* (possessors) whose plots of land would be directly affected by the park (the individual payments were estimated between 45,000 and 105,000 pesos annually, depending on size of plot and impact of development) and an annual contribution of 5 million pesos to a pension fund for comuneros over the age of seventy. The neighboring townships of El Zapote and Carrasquedo would each be allotted 250,000 pesos annually to offset the impact of the park on their lands and to help them reach their own development targets. The remaining, approximately 16.5 million pesos per year (about $1.25 million), would go into a community trust (*fideicomiso*) responsible for investing in social development projects like those outlined above.

Sergio saw establishing the trust as a vital aspect of Yansa's work. "We want to make sure the whole community sees the benefits of the project, not just the old men who traditionally run the comuna." To this end, Yansa outlined the composition of the trust such that at least seven of the thirteen trustee positions would have to be held by women and two by members of a youth forum (that Yansa had also helped to create). Only two trustees would be elected directly by the comuna, thus, in theory, dissolving the comuneros' ability to consolidate resources from the park for their own benefit.

Again Sergio said, "A very important concern is what happens to the revenue from the project. There are worries that it will create further inequality by making a few people rich at the expense of others. So this project, yes, is about building a wind park, but it's also about helping to strengthen and improve life in rural communities."

The Yansa model thus deliberately sought to decentralize traditional institutions of political authority like masculine domination of the bienes comunales. Sergio did not say this loudly, preferring to focus on the enfranchisement of "vulnerable populations" like women and youth instead. For the most part, however, these political goals seemed to be uncontroversial in our conversations with comuneros. Perhaps this was because a sufficient number of younger and female members of comuna families supported the initiative, perhaps because the provisions were lodged deeply enough in planning documents that were largely indecipherable to Ixtepecan campesinos, or perhaps because it was simply unthinkable that something as insubstantial as a wind park proposal could challenge the deep grooves of traditional institutions and relations of political authority.

When the "normal comuneros" spoke to us about the wind park, their interest tended to gravitate toward two issues: What kind of work might they receive during the construction phase of the project? And what kinds of rent payments they were likely to receive thereafter?

The *presidente* of the comuna, Baldomero Rosano, an old farmer, expressed the view of some that it remained unclear what Yansa was really offering them. People generally detested CFE, but their rental payments for running transmission lines over comuna land were clear or at least familiar. They brought in good money. "But with Sergio, it still isn't clear [what he will pay. It's] still not. . . . And the people don't accept that. Some are complaining. . . . They are waiting to hear the amount of the payment. What will Sergio pay? What will he pay?"

Uprising

The very next morning we receive our first real taste of the subtlety and precarity involved in exercising political authority in the isthmus. A meeting with a few dozen posesionarios went well the previous afternoon. Sergio was optimistic. There were questions of clarification, especially regarding the direct payment structure. But there were no major challenges to the Yansa proposal. This morning is the presentation to the full meeting of the comuna with the hope that they will ratify the proposal and, more importantly even, agree to send delegates to Mexico City to help Sergio put pressure on CFE to allow Yansa to bid on the open tender in October.

Sergio says, "CFE wants it to look like this proposal is just some foreign NGO's idea so they can marginalize us. We need to bring comuneros to DF to show them that there is an indigenous community behind this project.[13] That'll be tougher for the government to ignore."

When we arrive at the *comisariado* building, just down the street from Ixtepec's city hall, at first we think we have the meeting time wrong.[14] The street is filled with open-bed pickup trucks and a few tractors. Comuneros and a few comuneras are rapidly streaming out of the entrance, and the area in front is choked with people, some gathered in small groups around men who are holding forth in raised voices, some joking with each other, others having whispered conversations on the side. We wait until the exodus

FIGURE 1.1. Meeting of the Ixtepecan comuna

subsides and then cautiously elbow our way inside, greeting comuneros we recognize from previous meetings, hoping we will find someone who can explain to us what has happened.

We make it into the main auditorium. Filled with vividly blue chairs that are normally quite comfortable, right now the room is intense with emotion, talk, and bodies, the ceiling fans whirring the hot air around. We see Juan and Vicente, who shake their heads and gesture toward the podium at the front of the room, where Sergio stands looking a little stunned.

We ask him what happened, whether we missed the meeting.

"No, it never got started." Sergio goes on to explain how during the calling of the roll (a process that usually takes about two hours since more than eight hundred names are read aloud), a rival group of comuneros challenged the current leadership with not announcing a payment of 5,000 pesos for electricity transmission, effectively accusing them of having pocketed the money. "Many comuneros still mix up the wind park issue with electricity transmission rents more generally, so these guys were able to convince a lot of people that we couldn't talk about the wind park until this other issue had been resolved. And so they instigated this walkout."

The uprising had been staged by what Sergio describes as a "dodgy bloc," a group of squatters on comuna land (affiliated with the Ixtepecan mayor's family and the PRD party) who want to challenge the current comuna administration (members of the Confederación Nacional Campesina, CNC, the peasant organization of the PRI party) in an upcoming election.

Their dodginess did not stem solely from their squatting. The isthmus has a long, albeit contentious, tradition of land occupation dating back to the Mexican Revolution. At least in Ixtepec, we were told, if squatters occupied unused land and worked it, they could generally eventually count on a successful application to enter the comuna. But these particular squatters were also alleged to have connections to a mining company that was wishing to explore for gold in the area. In any event, it had been a skillfully staged sandbagging since many comuneros—including, in this case, all the main supporters of the wind park—tend to avoid the tedium of the roll call and show up for the meeting two hours late, when the real business begins.

Sergio is frustrated, calculating the fallout. "Honestly, the project is in jeopardy. Three years of work have gone into this, and everything depends on what happens in the next month." Not only will he have to convince CFE to change their rules for the tender but now he has to cancel a month's worth of meetings with social investors since he will not be able to produce an agreement letter from the comuna to show them. "I can't believe they would let a five-thousand-peso issue derail this when there are millions at stake for them."

By now we are standing in the foyer of the nearly empty comuna hall. The door to the comisariado's office opens, and we are quickly ushered in. The office is small and dusty but air-conditioned. A broken clock, an ancient desktop computer, last year's calendar, and especially the faded bienes comunales flag from 1956 all cry out for allochronic rendering.[15] But the energy in the room is utterly here and now.

There are ten of us packed in, and only the presidente, Baldomero, is seated, lividly angry and refusing to make eye contact. He drums his fingers on the green felt of the desk, occasionally shuffling through the member list in front of him.

The comuna secretary and treasurer dominate the discussion instead. They are in damage control mode, slightly defensive that they did not manage the confrontation more effectively.

"Is there any way we can count the meeting from yesterday as the official vote instead?"

"No, we didn't have a quorum of the general assembly."

The secretary says they know *los gritónes* (loudmouths) are "playing a game" but they still debate how best to defuse them.

"An official report from the treasurer?"

"They'll never accept that."

Some think these critics may be trying to delay action in the hopes of winning the November election in order to position themselves to get money from the Yansa deal. They discuss how and when to put out the *convocatoria* (call) for a new assembly meeting on the Yansa project. Two meetings would be best, owing to *la sicología de la gente* (the psychology of the people), the first to "calm the spirits," the second to make the vote happen. Another problem is that the asamblea has a lot of other business to cover. In the end, they decide they cannot schedule the first meeting until September 23, more than a month away.

Sergio flinches and raises the point that the CFE tender ends on October 4, leaving them very little time to act.

To reassure him somewhat, they agree to send two or three persons with him to Mexico City to request that CFE allow Yansa-Ixtepec to bid in the tender. However, they reject Sergio's proposition that they sign a statement in support of the park project as the comisariado because "that could easily be misinterpreted later as a sign that we had already grabbed some money from this project behind the back of the community."

Everyone is a little frayed by now. They find some catharsis in taking turns excoriating CFE for creating conditions that exclude community wind park projects in the first place.

Daniel, the comuna lawyer, says, "CFE wants to maintain their monopoly, that much is clear. If our wind farm is built, they're afraid it would help to expose all their corruption and excess costs they pass on to us. That [exposure] is what our screamers here actually want to happen, but they don't really know that's what they want."

Someone else ventures that CFE may even be encouraging its employees in the city to undermine and divide the comuna. "What CFE wants in the end is that everyone pays their high costs so that they are practically the only ones with wealth."

Another declares, "We'll take it to Congress if we have to!"

Rhetorical questions are reeled off, all aimed at the absent presence, CFE.

"Why aren't you letting us even participate, isn't that discriminatory?"

"Is this the best the Calderón government can do? Why isn't there a community wind farm of this size anywhere in Latin America?"

"Aren't you violating international as well as national laws?"

The hatred of CFE runs deep throughout the isthmus.[16] As elsewhere, so many of the desired conveniences of modern life (artificial light, televisions, air conditioning) make people dependent on those who provide electricity.[17] And that energopolitical dependency is only deepening as households add new appliances, even in the poorest parts of Mexico like the Isthmus of Tehuantepec, which in turn increase utility bills and financial precarity.[18]

Understanding the depth of negative affect requires thinking in terms of the clientelist political and economic relations that still predominate in this part of Mexico. In the logic of caciquismo, for example, one could at least expect some top-down redistribution of resources if one worked loyally for the cacique's network. The political party networks also offer ways of translating the available power of labor into comparatively scarce currency. In either model, a poor farmer could, for example, pledge votes and the presence of his body on blockades and in rallies in exchange for other kinds of support and resources. But CFE only exchanges electricity for currency. It accepts only the form of value that is already in such short supply.

The power company was thus a deceiver, we heard again and again. It brought light; promised progress, mobility, and modernity; and then made that progress contingent upon further impoverishment, blacking out dreams and further condemning *el pueblo* to hopeless marginality.[19] Part of the bitterness also had to do with the fact that CFE was, more than the army (whose physical presence was highly consolidated along highways and in urban centers), more than the parties (which were strongly rooted in local political networks), the routine infrastructure of translocal administrative power (e.g. "the state") in southern Mexico. And no matter whom we spoke with outside of CFE, it epitomized the negative reciprocity of that state: always taking, never giving.[20]

The secretary states flatly, "I'm for this wind project to break the corruption of CFE. They'll fight us because it is an example of how we are changing our life."

The meeting dissolves, and we catch Sergio for a few minutes before he leaves.

He remarks offhandedly (but in English) that the current group of comisariados is a little corrupt ("At first they used to suggest ways we could benefit them, but we just ignored that, and after a while they stopped asking"), but the group of challengers is "totally corrupt," so he still feels that Yansa is working with the right people.

But what, we ask him, about all that just happened.

"I'm really pissed off that they allowed such a weak argument to stop the meeting," he says. "Too many uninformed people dominated the discussion, and the people who were informed didn't show up until it was too late. Daniel, for example, could have explained the issues in a way that would have got us past this. Ultimately, they have to take responsibility and get their side of things in order. Yansa can't get involved in managing the internal politics of the comuna."

And yet the future of Ixtepecan wind power depends on the successful navigation of those politics. Bruised but unbowed, Sergio vows to keep working on the problem of CFE, hoping that his allies will be able to get the comuna solidified behind the project.

The Visionaries

Much of that responsibility would fall on the shoulders of Yansa's two closest allies, Daniel and Vicente, who belonged to the small group of comuneros who had been pursuing the possibility of a wind park even before Yansa arrived. Both are retired professionals in their midfifties, both deeply disturbed by the federal government's neglect of Ixtepec and the isthmus, and both very animated by issues of social justice and particularly indigenous rights. Both hoped to use wind power to change life in Ixtepec, making new opportunities available, especially for youth.

Vicente spent most of his life working as a chemical engineer for the Mexican parastatal oil company, Pemex, but returned to Ixtepec after his retirement and found himself depressed by the lack of possibilities for el pueblo. Vicente was also a philosopher, a historian, a collector of pre-Hispanic artefacts, a maker of *mezcal de nanche* that only became more sweetly potent as our conversations with him continued. He quoted Hermann Hesse and Karl Marx and told us that more than anything, Ixtepec needed opportunities to thwart the *indolencia* of large sections of the population.

"What is the problem with life here?" we asked him over a breakfast of nopales and queso oaxaqueño accompanied by mugs of thickly clotted atole.

"Are you familiar with the concept, 'neoliberal'?" he asked.

All too well, we replied, laughing.

"Well," he continued, "people here live on miracles. What has happened is that we live in a false economy. Why false? Because if you stop paying the teachers, nothing moves; the vendors suffer because there is no money circulating. And then if you stopped paying the soldiers, a week later the people would be dying from hunger. There'd be nothing left except water to drink. We produce nothing."

This was a theme that Vicente returned to again and again, Ixtepec as a dying agrarian community, one no longer able to produce enough food to sustain itself, captive to market forces far beyond their control.

He was not the only person to tell us, "There are only three jobs in Ixtepec: teacher, soldier, and market vendor." And only the first two jobs brought new resources into the community in the form of government wages. The vendors were less extractive than metabolic, sustaining themselves on the recirculation of meager fragments of capital.

Dalia, Vicente's wife, said she recalled perfectly the night Vicente came back from meeting with other comuneros and said he wanted a community wind park for Ixtepec. "He wanted development for the people so they would have employment and less violence, so they would have more happiness. I was so happy he had a positive plan."

Vicente offered us his own critique of alienated labor. He told us, "What we're always thinking about is *potencializar las capacidades*. To potentialize our capacities in a manner that can generate opportunity, opportunity that is more than one of these opportunities people have now. The plan is *potencialización*."

He explained, "It's the difference between *chamba* [job] and *trabajo* [work]. When I work, I'm working physically as well as intellectually. When I do a job, I'm doing it mechanically and don't care about the outcome because that makes no difference to me. For myself, I'm looking neither for jobs nor for work, I have a decent pension and children who won't have to depend on me. But I want opportunities for them. And I want opportunities for all."

Continuing, Vicente said, "This park could reach many, many people. This is why I'm putting basically all of my time into the project. Some people think Sergio's paying me to work on it. But he's not. There are some who just can't imagine that this project is without compensation [*sin dinero*]. The reason why I'm doing this work is that so many things are converging in this project, so many things that are affecting me."

For Daniel, affects were converging too. But for him, it was as much a matter of reckoning with pasts as imagining alternative futures. Daniel was the grandson of a rancher and a farmer; he fondly recalled visiting his grandfather's lands as a child. Now those lands belonged to him.

"But along that road," he said, "where there used to be ranch after ranch, there are now only four remaining. And you know how many people still live out there? Just one. Even I don't live there because I have so much going on in town."

In his retirement, he had taken up growing sorghum again, ten acres, and that taught him how much the infrastructure of irrigation had collapsed over the years as fewer and fewer farmers worked to maintain it. And other infrastructures were in decline as well, those that linked land and language and ethnicity. Daniel spoke *diidxazá* (Zapotec) in his youth even though he worked in Spanish throughout almost all his professional life. He was teaching his grandson diidxazá now, taking him out to places around the isthmus where diidxazá was still the primary spoken language.

Although Daniel was always quick to joke and chuckle, the undermining of indigenous rights in Mexico deeply troubled him. He saw the *núcleos agrarios* (agricultural centers) that formed bienes comunales, núcleos like the Ixtepecan asamblea, as having inalienable entitlement to their land. This was, for him, more than a moral claim, it was a black-and-white matter of both Mexican and international law.

"The Salinas government invented a procedure,"[21] he explained, "allegedly to regularize the núcleos agrarios in Mexico, but really for promoting the privatization of land. But this procedure addressed the ejidos, which were social forms, products of the Mexican Revolution and not comunidades de bienes comunales, which exist factually and rightfully as descendants of the first peoples of the Americas. Or, rather, they never really took the time to consider the difference between a comunidad and an ejido. Ejidos can now elect to privatize themselves, but not, as a matter of law, comunidades."[22]

Daniel understood this was a controversial claim, an appeal to an interpretation of the Mexican Constitution in which the rights of indigenous communities superseded those of the settler nation.

He spoke with admiration of the Zapatistas in nearby Chiapas. "Maybe they haven't done such a great job with development, but they've kept alive their right to *autonomía.*" His was, needless to say, not the dominant constitutional interpretation communicated to us by participants in wind development in the isthmus. But it was a crucial feature of Daniel's worldview. In his legal practice, Daniel had done significant work with núcleos agrarios across Oaxaca to help them better understand their constitutional rights.

"The issue," he said, "is that núcleos agrarios need to take advantage of their resources, but they don't have the money, let's say, to create a company of the kind that would be able to make productive or lucrative use of their resources. But here we have a possibility through the generation of electricity to create a company that would truly benefit the greatest possible number of members of a comunidad and not simply tiny segments of them."

He went on, "What we want is to be an example here in Mexico, but also in Latin America, of an indigenous community organizing itself, and collaborating directly, according to the agrarian legislation of Mexico, with a foundation, Yansa, to create an association, a community interest company, working in our collective interest."

He saw so many potential uses for the funds: "We could improve roads, construct reliable sources of water supply for our ranches, create educational opportunities for our children, pensions for the older comuneros, find real work for people."

When Daniel spoke of the significance of the Yansa-Ixtepec partnership, his Evangelical faith also came to inflect his language. It was a matter of coming up with more *justo* (just) forms of development. It was a matter of goodness (*bondad*). The partnership was itself *bondadosa*, an act of kindness. "No wind company anywhere in the world would come here and offer their expertise in exchange for our land and agree to share every peso we make equally. None."

Daniel's aversion to the dominant model of wind development found expression in the courageous pro bono legal work he did on behalf of the comuneros of San Dionisio del Mar to prevent the construction of the Mareña Renovables wind park. That story is told in the other volume of this duograph.

Daniel summarized his motivation quite clearly one morning over breakfast in Oaxaca City: "The real winners from the wind parks right now are in more or less this order, (1) the governor, (2) the politicians, (3) the companies that sell construction materials, (4) the private landholders, (5) the unions that get the construction contracts, and (6) the scraps for everyone else."

When we arrive back in Ixtepec in late September for the next asamblea, we find that Daniel, Vicente, and the comisariados have done their work well. Even the mercurial leader of the gritónes, Isaias, has been brought on board. Internal factional politics aside, it seems he shares an intense distaste for both CFE and the international wind developers who are at work elsewhere in the isthmus.

But the situation has become more complicated for other reasons. Sergio has just learned that CFE refused Yansa-Ixtepec's application to bid in the October tender. The rationale was that they lacked the requisite cash in the bank and letters of credit to produce a credible bid.

Sergio is disappointed but not surprised. "Of course, nowhere in Mexican tender law is there any requirement for this kind of money in the bank. CFE is just creating new laws itself." If lawmaking belongs to the domain of political power in the Lockean sense, CFE has revealed energopower's capacity to colonize that domain for its own convenience.

The assembly is a long and complex one. Clarifying the question of the payments for the electrical towers takes a long time; many voices weigh in. The comisariado struggles to maintain order, and the discussion frequently dissolves into shouting. Some chide the leadership that they are undertaking a "public function" that requires full transparency. One senses that the oratorical groundwork is being laid for future political campaigns.

Then there is an unexpected interruption from an older woman who pleads for the restitution of land she vacated over thirty years before when her husband, a soldier, was stationed away from Ixtepec. After five years, unoccupied land reverts to communal use, and her land, buildings and all, has been taken over by someone else. But now that her husband has died and her children are grown, she wants to return home to Ixtepec. By the end of her plea, she is in tears, but many speak against her claim ("Thirty years gone!"). In the end, it is agreed that she can be given a different plot of land. Almost all the important arguments concern land. Land is the basis of the comunidad, it is patrimony, and at least in this auditorium, it is the ultimate source of value. Terroir—here, literally "soil and its capacities"—propels politics.

The asamblea are tired, distracted, and cranky by the time Sergio is allowed to make the announcement that CFE has decided not to allow them to bid in the tender. Attention refocuses quickly; the auditorium goes still but then begins to rumble as Sergio walks them through this latest outrage. How CFE has excluded the comunidad from even competing to access a substation that lies *on their own land*. How CFE has already given away the rest of the access to six foreign companies ("four of them Spanish"). How it is demanding $7 million up front and 548 million pesos in the bank to participate in the competition. How they want the tender winner to also pay to prepare the land for a new wind park in nearby La Mata because it saves CFE money.

"In accordance with the laws of Mexico, everything that CFE is doing is illegal, violating the constitution, violating agrarian rights, the internationally recognized rights of indigenous communities and campesinos, violating even its own norms."

Sergio asks the asamblea how they want to respond, and someone yells, "*¡Hay que demandarlos!*" (Let's sue them!)

The crowd stirs further as Sergio discusses some of the forms of collective legal action that could be taken, including *amparo*, a provision of Mexican law that allows individuals and groups to seek protection from unconstitutional abuses of the government to defend civil rights as outlined in the first twenty-nine articles of the Mexican Constitution.

"*¡Lo vamos a hacer!*" (We're going to do it!), someone else shouts.

Isaias then stands up and asks permission to speak. Hardly pausing, he launches into a partly handwritten, partly improvised, thoroughly firebrand manifesto criticizing the authorities of CFE, the government, politicians of all levels, as "*antipatrióticas*" who are doing dirty business with foreigners, heaping abuses on the pueblos of the istmo. Here is a taste of it:

The Mexican government doesn't follow or recognize the international treaties because the gringos, sorry, the gachupines [Spaniards] are giving them money hand over fist, starting with the secretary of the interior, who gives them concessions, permission, and authorizations that streamline the whole process. They jump as quickly as the Spaniards want them to. They're mad because they've come to loot us and we've taken the time we need as a community [*pueblo*] to analyze our situation and build our rescue plan [*propuesta de salvamento*], to address the sharpening social, cultural, political, and economic crisis generated by the devastating policies of the state that results in violations of the human rights of people in the region and in the country. Nowadays governments are used by violent capital who buy the wills of authorities who then threaten or do not hear what the compañeros tell them about their needs. They assault, deceive, and corrupt the people. Compañeros, we are in a situation that has caused conflicts inside and outside of our communities. These projects mean the dispossession and destruction of our environment, flora and fauna, all with the complicity of the federal government and the municipalities too.

Isaias switches into diidxazá, shouting, "We won't be misled mules, each animal running in its own direction!"

This provokes widespread laughter, whoops, and hollers of recognition, "*¡Paleros! ¡Paleros! ¡Somos Paleros!*" (Leaders! Leaders! We are leaders!)

And finally, bringing it home, he offers a proposal: "Either we have a community wind park here or no park at all."

Rounds of assent follow.

People start to get up and drain out the back of the auditorium. But Sergio is still pensive, a little bewildered perhaps by all the shouting. And, everything is on the line.

He calls out to the retreating asamblea, his right hand raised in a tentative thumbs-up, "So you agree to adopt compañero Isaias's resolution?"

"Yes! Yes! Yes! Yes!"

Squatter Grid

Daniel, Sergio, and others in the asamblea had begun planning legal strategy long before CFE actually turned them away. But the ten days following the asamblea were feverish as they developed an effective case against CFE. We accompanied Daniel and Sergio to Oaxaca City to consult with an indigenous law specialist at the Oaxacan Ministry for Indigenous Affairs. The specialist advised them that a complaint focused on being denied access to infrastructure was weaker than one focused on the subversion of their constitutional right to use their land the way they saw fit. There were other legal options, of course, but he confirmed Daniel and Sergio's sense that amparo was the right strategy. "It's faster, it'll put pressure on CFE to deal with you. *El amparo tiene mas cajones.*" (The amparo has bigger balls.)

In his role as the asamblea's lawyer, Daniel moved quickly to draft two amparo complaints against CFE on behalf of the comunidad, one addressing each of the public tenders that CFE was offering at the time, Sureste I Fase I (200 megawatts) and Sureste I Fase II (100 megawatts). Daniel submitted these complaints to the Sixth District Federal Court in Salina Cruz, naming the three members of the comisariado as the complainants. In the complaints, Daniel stated that CFE's actions had violated five articles of the Mexican Constitution, one article of the American Convention on Human Rights, eight articles of the Agrarian Law, and five articles of the Indigenous and Tribal Peoples Convention of 1989.[23]

In a video interview with us for purposes of documenting the legal struggle, Daniel summarized the asamblea's legal argument as follows,

> The CFE and other authorities approached the asamblea of the bienes comunales with a request to *study the possibility* of building a new substation, the largest in Latin America, for the new wind parks. And for that they needed forty-two hectares of land. Because there was some hope that this could generate jobs, of which there are few opportunities here, the asamblea gave CFE the authorization to do *the studies* for the viability of constructing this substation. But the CFE didn't just do the studies. It actually went ahead and occupied the land and began to construct the station even though it still didn't have that authorization. They never presented the results of those studies to the asamblea. And furthermore they deceived the community in their presentation by saying we would have the opportunity to build a wind park. But then, when the opportunity arose to compete for access to that substation, CFE closed the door and said *"no vas a entrar"* [you may not enter] because you don't have money. So this was a case of exploitation and theft [*despojo*] on two grounds. First, the illegal occupation of our land by the substation and, second, by refusing us the right to even compete for access to the substation.

In addition to CFE, the complaint targeted the Mexican federal environmental ministry, SEMARNAT, for having granted the CFE the right to *cambio del uso del suelo* (change of land use) on the forty-two hectares based on an environmental study that was itself performed, according to the complaint, on an illegal basis.

It was an argument with a factual basis in that CFE had apparently never signed an actual contract with the Ixtepecan assembly concerning the plan to build the substation. There was also no formal expropriation of the forty-two hectares by the state. The electric company had simply, as Daniel put it, occupied the land and begun building. A squatter substation.

This might seem like a startling oversight. However, this was a context in which the agencies of the Mexican federal government and their corporate partners in wind development seemed generally uninformed about, or simply impatient with, bienes comunales decision-making procedures. Any issue involving cambio del uso del suelo typically required at least two assemblies to be convened to discuss the issue, and then it had to produce, in the presence of a notary, a majority vote of all current registered comuneros.[24] Dubious procedures and suspect notaries became critical issues in the Mareña case as well. And more than this, the image of CFE occupying indigenous land with its squatter grid was powerful and paradoxical, a cunning use of *mestizaje* nationalism that appeared to contradict the settler sovereignty of the Constitution.[25]

Still, Daniel confided in us, he was doubtful that any legal strategy would produce the outcome they wanted in the long term. "Our best hope is that it forces some kind of political compromise."

The next week brought an important legal victory for Yansa-Ixtepec. A judge in Salina Cruz issued an order of suspension to CFE on October 11, 2012, on the Sureste I Fase I project, demanding that the tender process cease while the evidence for the comunidad's amparo claim was assessed. This was the first judicial injunction ever issued concerning a wind park in Mexico.

Then, the next week, Ixtepec's representative in the federal chamber of deputies, Rosendo Serrano Toledo, the brother of Ixtepec's mayor, announced his support of the Yansa project, arguing that the current model of wind power development "feeds social conflict . . . while, by contrast, the community wind model guarantees the control of *pueblos originarios* over the natural resources of their territories and gives the communities a central stake in the projects, allowing wind energy to become a motor of prosperity."[26]

These events did not go unnoticed in the broader field of Oaxacan wind power. We happened to have coffee with a senior manager of one of the

FIGURE 1.2. Ixtepec Potencia substation

major Spanish wind development companies about a week later, and he launched into an unsolicited tirade against Yansa, telling us they "were ripping people off" since their numbers did not make sense. "How can they offer people 50 percent of their profits from electricity sale when the market rate in the isthmus is closer to 2 percent? Something is very fishy about that."

But what happened next—or rather, what did not happen—was an early lesson for us in the material contingencies of political power, specifically in how an energy-supplying parastatal could operate as its own sovereign "state." Simply ignoring the judge's suspension order, CFE proceeded with both tender processes.

Daniel and Sergio attended a meeting at CFE in Mexico City on October 30 in which the several companies that had been allowed to bid in the tender presented their economic plans (including the critical data point of the production price of electricity per kilowatt hour). At the beginning of the meeting, Daniel stood up and waved a copy of the judge's suspension order, saying that the tender could not proceed. But he was told by the presiding senior CFE official that the company had not been informed of any suspension order and that they would certainly not accept a hand-delivered

FIGURE 1.3. President Felipe Calderón at Piedra Larga

document from him. The tender meeting proceeded with no recognition in the minutes of Daniel's challenge.

Electricity, capital, and biopower were working in smooth concert in Mexico that day. Far away in the isthmus, President Calderón helicoptered in to a carefully orchestrated inauguration of the Piedra Larga wind park, an industrial self-supply project for the Mexican baked goods giant Bimbo located near Santo Domingo Ingenio. A caravan of several dozen protesters was intercepted miles away by a blockade that had been organized by a group of supporters affiliated with the construction unions.

Calderón stepped out of his helicopter and was greeted only by rows of CFE employees in matching crisp white shirts and blue CFE baseball caps and by radiant flower-adorned women in all the bright colors of Zapotec dress. He spoke of his passion for renewable energy, of extreme weather events and the reality of climate change, of the importance of addressing climate change for all humanity, of the job creation the parks would bring the isthmus, of the economic revenue from the land rents, of the new capital investments and infrastructure projects that would follow. He recalled his first visit to the isthmus and how he had wondered why "we had only these tiny parks built by CFE where the machines looked like scarcely more than fans. . . . How can Oaxaca have only two megawatts of production when it is one of the most important areas in the world in which to generate wind? . . . Now Oaxaca has converted that promise into a reality, it has become a protagonist of electricity generation for our country."

As Calderón walked back to his helicopter, he nearly stepped on a tarantula. He paused and took a photo like a curious tourist. The trailing CFE engineers took pictures of their president stooping down, holding his camera at arm's length. And then Calderón returned to the sky, having said nothing at all about the possibility of community wind power.

High Voltage

Daniel filed a new complaint in Salina Cruz, stating that CFE had violated the suspension order for Sureste I Fase I. The judge issued an investigation of CFE's failure to comply. A different judge, meanwhile, turned down the amparo against Sureste I Fase II on the grounds that indigenous communities had no legal interest in the production of electricity. Daniel moved to appeal, but a few weeks later, on November 22, CFE issued the announcement that it had completed its tender process for Sureste I Fase II and had awarded the 100 megawatts of grid connection that Yansa-Ixtepec had hoped for to Impulsora Nacional de Electridad (INE), a filial of the Italian company ENEL Green Power. Their bid of $0.0412 per kilowatt-hour had come in below competing bids from Acciona, Aldesa, Elecnor, Iberdrola, and Isolux Corsan.

But through some of alchemy of pressures—from the judge, the federal deputy, the new lawyers Sergio had recruited to the cause in Mexico City— in December, CFE finally began to respond to the amparo for Sureste I Fase I, and the case entered the discovery phase.

On one car ride to Salina Cruz to submit evidence, Daniel joked, "I say we're in the evidence phase, but CFE isn't. They're just denying that any evidence exists."

We accompanied Daniel on other trips around the isthmus in the weeks that followed, but our attention to the Ixtepecan amparo was being torn away by the gathering maelstrom surrounding the Mareña Renovables project.

When we were able to refocus again on Ixtepec, we thought it important to talk to someone at CFE about the controversy surrounding the substation. We began to haunt the main CFE administrative center in Juchitán, sitting in a lot of waiting rooms, talking to staff, cold-calling various directors and subdirectors, all with little success. Eventually we were able to set up an interview with an *ingeniero* (engineer) named Nucamendi, the head of the Transmission Division for the region and manager of Ixtepec Potencia, the region's major CFE substation.

On a February day thick with smoke from the seasonal brush fires surrounding Juchitán, we showed up at the Transmission Division headquarters, a bustling modern office hidden away on the second floor of what looked from the outside like an abandoned factory. Nucamendi greeted us, shaking his head. He spoke kindly but said that he had requested permission from his *jefe* in Chiapas to give the interview and been turned down. "Your questions are sensitive. I would like to talk to you but *no se puede.*"

He thought an engineer named Cervantes over in the Distribution Division might still be allowed to speak. But we didn't make it any farther than Cervantes's secretary, who told us firmly that "CFE has nothing to do with renewable energy."

"But what about the grid and power lines?"

"No, we have nothing to do with energy, others are supplying it."

"But without CFE, there's no electricity, right? You are the distributor."

"That's a different way of thinking about it."

The company's way of thinking about themselves appeared to be as a purely technopolitical service infrastructure. But we marveled in the weeks that passed over their exquisite skill in thwarting our every effort to speak to someone, never by saying no, but always suggesting yet another someone who was always in a meeting, out of town, or visiting an installation somewhere. It became a weary running joke.

Eventually, we ended up with the director of the two CFE-run wind parks (La Venta I and La Venta II) a friendly ingeniero named José Manuel Benitez Lopez. Somehow he was far enough removed from the conflict that he was never warned not to speak with us. His office was a small bungalow down a cul-de-sac in the company's main campus. On the shelf behind his desk, a small model of a Spanish galleon, a glass model of a large HV electrical tower, and a white plastic model of a wind turbine were casually (but to us, provocatively) juxtaposed. When we asked about what was happening with the amparo in Ixtepec, he simply threw up his hands and said he had no idea. It wasn't his department.

But he told us more about the substation from CFE's perspective. Ixtepec Potencia was the biggest substation in all Mexico, maybe even in all Latin America. "This place takes up forty hectares. It's huge. *Un monstruo, eso.*" (It's a monster.) Part of the reason for its size was that it was also one of the most technologically advanced substations that CFE had ever constructed. It included two transformer banks of 440/115 kilovolts and three of 400/230 kilovolts, with a total capacity of 1.875 million volt-amperes. It also brought 400 kilovolt high-voltage power lines, switchgears, and busbars into the

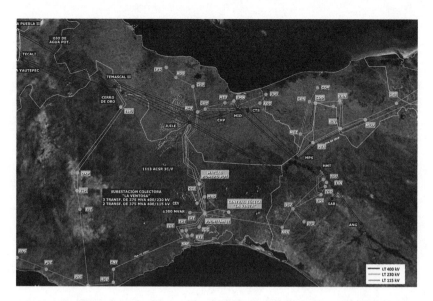

FIGURE 1.4. Map of the southern Mexican grid showing wind parks

isthmus for the first time. But most of the power would be used elsewhere, north toward the substation at Juile and then on to the northwest toward Mexico City.[27] To help manage and regulate grid voltage, CFE had also installed a state-of-the-art Static Var Compensator (SVC) at Ixtepec Potencia, which promised near-instantaneous response to changes in system voltage.[28]

"It's sort of like a battery, but not really a battery," Benitez explained. "But it's very important since it is connected to the wind parks and the intermittency of wind energy is challenging from the point of view of grid management."

The substation at Ixtepec was thus a first foray of CFE into creating what has been termed a "smart grid," able in this instance to incorporate fluctuating electricity sourced from renewable energy smoothly with baseload thermoelectric supply. In southern Mexico, which produces twice the electricity it consumes, there is no real fear that on less windy days there will be power outages.

He sighed, "We can basically shift over to the hydroelectric power from Chiapas at any time. The people don't care where the electricity comes from. They just want the light on when they hit the switch."

Benitez also characterized Ixtepec Potencia as an infrastructural investment that belonged entirely to CFE, a key node in the largest federal investment in the isthmus since the Transisthmus Highway was built in the early 1950s. But we heard whispers throughout government and industry that CFE had made a deal with private wind developers to cofinance the construction of the substation, possibly in exchange for preferential grid access. If true, this would have made the "open season" (*temporada abierta*) of bids and contracts into yet another deceptive CFE performance with the winners predecided.[29] No one we spoke to at CFE in Mexico City would confirm this arrangement, but in the isthmus, the rumor was widespread. The substation, like other Ixtepecan squatters perhaps, was simply holding its ground, working, and waiting for eventual recognition and acceptance.

We later confirmed that developers were responsible for doing the gridwork of getting the electricity from the point of production to the high-voltage transformers in Ixtepec. Because of the parks, the grid was expanding, its metal tendrils creeping across the isthmus, humming with wind-generated current.

Twenty years from now, Benitez explained, the wind parks, including all the new transmission infrastructure, would revert to ownership by CFE. "But of course," he chuckled, "the life span of a turbine is something like twenty years."

Tristepec

"Anyway," Raul Mena shrugs, "even if it were to happen, this park project would only benefit the comuneros anyway. There's no community interest in this project. Ixtepec has maybe twenty-five or thirty thousand people living here, and the comuneros are only a small part of that, less than a thousand. It's *herencia* [legacy], lineages of *ancianos*, old men from old families. When the substation was built, all the construction work went to the comunero families. But the majority of families in Ixtepec don't have land because we arrived later. So we don't care about this wind park. It won't bring us anything."

Raul's sister, who works in city hall, nods along with him, "They're not exactly secret over at the bienes comunales, but it's limited who participates. *Ellos son cerrados.*" (They are closed off.)

Raul belongs to the ranks of the educated and underemployed young adults of Ixtepec. He has not migrated to the North or sought to leave the country. But with proud resignation and a little bitterness, he lets it be known that there is little in Ixtepec—"Tristepec" he jokes—for people like him. Again, the flux of infrastructure has played its part. In 1907, when Porfirio Díaz's railway to Guatemala was built through Ixtepec, the city became a hub for traveling merchants, commercial interests, and immigrants (gringo, European, Arab, Asian) seeking markets, labor, and influence in the isthmus. But then in the 1940s, a powerful mayor in Juchitán, Heliodoro Charis, made sure that the Trans-American Highway went through Juchitán instead of Ixtepec, and that established Juchitán as the commercial and political center of the region once again (see chapter 5). Ever since, Ixtepec has been in decline. The railroad endured, but it no longer brought promise to Ixtepec. It brought only a train known as *la bestia* (the beast), which was famous in the Mexican media for its transport of thousands of illegal migrants from Central America into Mexico each year—new prey for the cartels—thus, it was sometimes called *el tren de la muerte* (the train of death). A few Central Americans stay in Ixtepec, mostly in homemade shelters near their tracks, their possessions in small backpacks. In Juchitán, they are blamed for most of the street crime. But that is not Raul's feeling about the migrants: "They don't come looking for trouble. And there is nothing for them here either. Mostly they move on."

Neither Raul nor his friend Isabel, a local teacher, believes that the wind park offers a way forward either. They are under the impression that Yansa is a for-profit investor like any other and assume that whatever resources the project brings in will be retained by the comuneros.

It troubles Isabel that there is no plan to reactivate agriculture in the region. "You can't eat money; it does nothing for productivity." There have been shortages of corn in the region already because, "the farmers are getting lazier and growing easier crops like sorghum and peanuts instead." Investment in agriculture would be better than building wind parks. "Agriculture is a general benefit," she says. Or perhaps education.

But Raul points out that Ixtepec already has a lot of schools, so many that it imports students. The problem is not too little education; it is "the lack of employment."

Isabel notes that "our professionals here are the children of campesinos, people who worked hard to advance themselves." But now there is no local industry and few white-collar jobs.

Raul sighs grimly. "All that people dream of here is to become a teacher, work for Pemex, the army, or CFE." Raul says that Ixtepec is *malinchista*,[30] that it has been that way for decades, and that the race for wind parks is merely more desperate imitation of a way of life that is not istmeño. It is "like the cement we build with, which is wrong for the climate and detracts from the beauty of the town." Then he waves his hand, "But still, if the park brings infrastructure here, then that's fine, even if there is no plan for the community." Raul's sense of "community" is clearly very different from that of the Ixtepecan comuneros, signaling how deeply alienated some citizens of the Ixtepecan municipio feel from the agrarian commons.

Adán Cortez works for CFE but in the office where people pay their bills. He has a comfortable office with air conditioning. He comments, "The people from Yansa speak well. They say everyone is going to benefit. But those who own the land are only a segment of the population. All thirty thousand of us don't get to vote in their asamblea. I grew up here. My grandfather worked for the railroad, and my father as a journalist. But we have no land."

So he is against the project?

"No. Look, there is just a lot of disinformation and mistrust. The people who know something are saying let's wait and see. Let's see if they do something concrete, maybe create a computer learning center or a library for the youth, maybe when they rent the machinery to build all this, maybe if we come to them and say, 'Could we just borrow that to do this little job at my house?' That sort of thing."

Dreams hang in the air in Tristepec. Talk centers on aspirations for more work, for development, for greater access to *recursos* (resources). Often, outside the comuna, renewing land and agriculture seems the most promising way forward. Even among the Ixtepecan professional classes, agrarian ideology exerts a powerful force—desire looking backward, as is its way—with many seeing future prosperity through a return to the earth. Yet the communal land tenure system dating back to the Mexican Revolution is actually part of the problem. The urban middle-class Ixtepecans say that the land is controlled by the ancianos in the comuna, people without education and imagination.

We ask Raul to whom the Ixtepecan wind belongs, and he says, "To us! To all of us here! It's not like *terreno* [land] but it is partly of the *tierra* [earth]." We have the same slippage in English between "land" and "earth," undoubtedly owed to our own agrarian legacies past. But Raul is making a more specific point—the wind is *not* like the land of the comuna; it is not something that can belong to just some Ixtepecans. The wind carries earth within itself, lifting fine particulate matter from the bienes comunales and scattering it through the city. The wind hints at the possibility of a broader redistributive commons, echoing the yearning of those outside the comuna for a future enabled by local soil.

One evening Vicente kindly arranged a focus group of sorts with several of his neighbors at his ranch house outside of town. Daniel dropped us off, and as we prepared coffee and waited for people to drop by, Vicente showed us, with obvious pride, the trees he was carefully nurturing on his property one by one: mango, coco, platano, moro, xiguela, papaya, zapote, almendra, and granata. Dozens of saplings were growing in plastic jugs. We tasted some of the fruits, intensely flavorful and very dry, like the land before the rains come. There was some discussion of how to remove a mating pair of iguanas that had settled under the corrugated roof. Vicente showed us some ancient artifacts he had recovered nearby, including a carved stone that he said was of a kind that was often tied to a stake in front of a chieftain's house.

As the neighbors arrived, Daniel left, and Vicente made himself scarce with the excuse that he needed to water his plants. He explained later that this was to give his neighbors privacy to speak openly since they knew he supported the project. The rest—Antonio, Florentino, David, and María Isabel—settled into the warmth of late-afternoon sun, drinking coffee, sharing bread made from rice flour and cinnamon. They were from comunero families that participated in farming and ranching to survive. They felt that Yansa was a good organization but that the situation with CFE was *muy delicado* (very delicate). "CFE may be very strong, but they are working against the people of Mexico."

The presence of CFE was real, the monster on their land. The wind park was elusive, speculative. No one seemed certain it would ever come. We talked about the stories associated with the land around Ixtepec and they told us of a *laguna encantada* (enchanted lake) just three kilometers from where we were sitting, which used to brim with magical entities dating back to pre-Hispanic times: fish and multicolored horses and even a golden comb that would lure the desirous into its waters. Now the lake is a dried basin, with many of its spirits having fled eastward, "toward Guatemala."

We asked them whether there were any legends concerning the wind in Ixtepec. They looked at each other searchingly for a few minutes, but no one could think of a story about the wind. There were no stories that could hold the wind to the land.

And then, his green eyes wide, David said, "Well, out east near La Venta, they say the wind can blow over trucks."

Endless Impasse
————

The Sureste I Fase I tender was cancelled by CFE on March 6, 2013. They gave no explanation, admitted no wrongdoing, suggested no reconciliation. Nothing was really solved as far as Yansa-Ixtepec was concerned. It was still unclear who would eventually get those last precious 200 megawatts of access to Ixtepec Potencia.

In May we met with one of the lawyers in Mexico City who was working on the case. We asked him to tell us what CFE's response to the amparo had been.

He said, "Their response has been nothing really. [It has been] completely technical, whether an amparo can be issued on these grounds or not. I think they are scared of touching the human rights issue, they don't want to go anywhere near that."

When we met Sergio for lunch the last time in Mexico City, he remained optimistic that some sort of political resolution could still be found. He recalled that, early on, there had been informal talk of granting Ixtepec a special exemption for a community park even if it did not fit the dominant development paradigm. "Even if it was only a gesture, it would still be a way of enfranchising Mexico's indigenous communities more substantially within the process of renewable energy development."

Recently, he had been speaking with officials in the ministry of energy, SENER, and found they were sympathetic to him and had their own (quiet) doubts about how wind development was unfolding in the isthmus, particularly in the wake of the spectacular failure of the Mareña Renovables project. Mareña gave Sergio an excellent opportunity to promote Yansa-Ixtepec as an alternative to the dominant *autoabastecimiento* model (industrial self-supply, see chapter 2).

There were smart people at SENER who understood, he said, that they needed to develop a new model of community engagement. The problem remained CFE: "They are just thinking about this as an engineering problem; the social issues don't matter to them." And then, of course, there were the clout and momentum of the foreign companies seeking to make money from the winds of the isthmus and the Mexican politicians and regulators who held private-public partnership as an article of faith for the future of Mexican electricity generation.

We had a meeting with a senior manager of one of the transnational firms investing in istmeño wind parks shortly before we left Mexico. When we mentioned that Yansa-Ixtepec was one of our case studies, he laughed.

"Oh, Sergio, he's a promoter, promoting things that are difficult to achieve. He's caused a lot of noise over at CFE. There are people like that who like to promote things but don't have the resources or the know-how to achieve anything."

We found that Sergio was always portrayed in the business and investment community (and often among government officials) as either an opaque political operator or, as he himself suspected, a madman, the Quix-otic counterpart to his sober industrialist Spanish countrymen, pursuing mirages in the light and heat of southern Mexico. Time will tell.

When we left Mexico in July 2013, the amparo concerning Sureste I Fase I was not yet resolved.[31] At the time, Daniel told us he was optimistic that a judgment would be imminent. In June 2014 the case remained unresolved. "One more month," Daniel told us. "There's no turning back."

After hearing little news in 2015, we spoke to Sergio again by phone in July 2016. He reported some dramatic twists and turns but no end in sight to the impasse. Having still never formally responded to the community wind park proposal, CFE returned to the Ixtepecan comuna in the summer of 2015 with a request to lease five thousand hectares of the Ixtepecan bienes comu-nales to build two new private wind parks totaling 585 megawatts as well as new transmission capacity to evacuate electricity.

Sergio thought this was an indication that CFE was planning an increase of possibly three times the number of istmeño wind farms in the future. The company "had people on the ground for months, trying to convince comune-ros one by one, trying to get them to sign." But Sergio observed that this kind of influence peddling was illegal under the terms of the new energy-reform law, with its guarantees for informed consent and social impact assessment such that "no company is allowed to enter into any kind of contract without a social impact evaluation conducted together *with* the community. Only *then* can there be a land lease." Sergio felt certain that such an evaluation process would resolve in favor of the community wind park as it had in the past. "But they didn't do any evaluation with the community. They were just trying to get contracts like always."

Events took a still-darker turn in the fall of 2015. After a failed attempt to circumvent the authority of the comisariado and to have the five thousand hectares transferred directly to CFE's control by the Procuraduría Agragria,[32] CFE backed a group of comuneros affiliated with the CNC (the Confeder-ación Nacional Campesina, an organization of ejidatarios and comuneros affiliated with the PRI party), whom Sergio viewed as "highly corrupt," in the November 2015 comisariado elections. They won the election in part through the installation of an irregular system of individual voting in place of a com-munal assembly decision and in part through the disorganization of the anti-CFE camp (which ran four different sets of candidates).

When CFE returned again to make a pitch for their five thousand hect-ares, one of the spokespeople let it slip that CFE had, in its view, already completed the social impact evaluation for the project. After much legal wrangling, those in favor of the community wind park secured documents that CFE had sent to SENER, asking that the normal consultation process be waived since, according to CFE's argument, although Ixtepec was indigenous, it lacked the "*sujeto social colectivo*" (collective social subject) that would make such consultation possible and necessary. The request was approved.

Sergio explained that, in essence, Ixtepec's energopolitical value had become such that the project could not be allowed to collapse: "Ixtepec has become too critical for them in terms of the infrastructure for this plan to fail. They are planning to build three gigawatts of new high-voltage lines directly from Ixtepec to the Valley of Mexico. Still, the majority of comuneros find the new leadership very suspect, and CFE has not yet been able to secure a single hectare of land."

Sergio fumed about the lack of transparency in this "reformed" process of securing land leases for large energy projects. "It's completely in the hands of the government and the companies, and as always, the communities are left out. If CFE hadn't accidentally admitted that a social impact evaluation had taken place, we'd never have found out what they were up to."

Were he and his allies tempted to give up on the Yansa-Ixtepec project?

"Not at all," he assured us. "With at least a doubling of the number of wind parks in the isthmus coming, this has become a regional issue. And other communities have already expressed interest in the community wind model. So we're working to bring all these communities together across the isthmus to prepare a collective amparo application to challenge not so much CFE's flawed social impact evaluation as the ruling from SENER to allow CFE to proceed with the projects. This is our front line now, organizing regionally to challenge the present course of wind development."

And there the story rests for now. The future of aeolian politics in Ixtepec remains unknown, with the community partnership model brightened by occasional rays of hope but beset overall by lengthening shadows of doubt.

What do we learn from the case of Yansa-Ixtepec and the failure, for the moment, of community wind power in Mexico? Yansa-Ixtepec gives us a glimpse of how new energopolitical potentialities are struggling to come into being in the Isthmus of Tehuantepec (and not only there).[33] Yansa-Ixtepec follows the charge of Scheerian thinking, seeking to harness renewable energy sources to transform and improve the social and political conditions of humanity, to bring justice and empowerment to long-marginalized indigenous communities in the postcolonial world. But instead of finding the high-voltage infrastructure of national enablement, Yansa-Ixtepec's vision has been kept off grid in more ways than one.

This chapter has mapped the forces that have intervened to block a new developmental model for wind power in southern Mexico: a policy regime thirsty for foreign capital and investment (but only as guaranteed and managed by "reliable" corporate partners); the desire of international capital and its agents to propagate themselves in a lucrative and low-risk way; the energopolitical authority and vast bureaucracy of a parastatal electricity monopoly; the enduring vitality of colonial clientelist legacies of political power, agrarian legacies, and ideologies that represent land as the true source and optimal form of value; a regional history of marginality to the biopolitical institutions and imaginations originating in Mexico City; racial and ethnological discourses of "ungovernability"; and, not least, the question of what really counts as "community" in a place like Ixtepec, where the lands of a núcleo agrario have come to include a municipality of former and current migrants, an army base, scores of pawn shops and market vendors, unemployed intellectuals, and even a substation.

The case of Ixtepec captures the new horizons of energos and bios that are becoming thinkable today as well as the resilience of extant infrastructures, institutions, and lifeways that inject paralytic compounds into the political imagination, making the nonrepetition of the present appear to be a fantastic unrealistic desire. Ixtepec shows us the oscillating life-and-death drives of aeolian political imagination: images of revolutionary breakthroughs and alternative futures coexisting with repetitions of habit and tradition. It gives us a glimpse of the turbulence that gathers along the edges of flows of enabling pouvoir as they swirl around sites of rapid energy development. And because it is still not clear whether Yansa-Ixtepec will never be or whether it is simply not yet, it even captures something of the paradoxical quality of life and the experience of time in the Anthropocene. We do not know whether what passes for the inevitability of the moment will endure or whether a revolutionary transformation of the contemporary is yet possible.

From the not yet we move next to an intense situation of the right now—to La Ventosa, an istmeño community that, in the space of our fieldwork between 2009 and 2013, came to be almost entirely encircled by wind turbines.

2. La Ventosa

July 21, 2009. A call from Fernando rattles us awake shortly before four in the morning. He's been up all night drinking red wine "with the Spaniards" and now wants to come right by and pick us up. Something important is happening in the isthmus. This call is a complete surprise; we are disoriented and, frankly, hungover ourselves from the mezcal that we thought was celebrating the end of our first research trip to Oaxaca. We beg for more time, but Fernando insists that it is now or never. So at 5:30 AM we are huddled downstairs, waiting in the dark. A midsize Ford pulls up not long thereafter. Fernando and his driver welcome us with more cheer than seems acceptable under the circumstances. Fernando Mimiaga Sosa, the director of sustainable energy and strategic projects for the state government of Oaxaca, is tireless in his promotion of wind power.

Compressed into the back of his car, we ask what is happening.

"It's an issue with the Iberdrola project; there are some doubts about payments having to do with *afectationes* (impacts), and so I need to go there to meet with them together with another official—you know [he laughs] I have no time for myself any more. I'm always on the run. I'm up all night with one group of Spaniards and then I get a call that I have to resolve this conflict between another Spanish company and the *dueños de la tierra* (owners of the land)."

The highway from Oaxaca City to the Isthmus of Tehuantepec is the Carretera Panamerica, and it is spectacular. Once we leave Oaxaca Valley, incredible vista stacks upon incredible vista as we wind our way through the Sierra Madre down to Jalapa del Marqués and Tehuantepec. Agave cacti grow wild, thick, and dangerously beautiful on the steep mountainsides. The *curvas* (switchbacks) of the two-lane road are nauseating until we get used to them. Still feeling the flush of wine, Fernando closes his eyes and sings along to the radio.

But today seems a bit too fringed with possibility for rest. And despite the alleged urgency of getting on the road, there are detours and side adventures along the way. We stop by Fernando's favorite *mezcalero*, driving down a dirt road to a cinderblock hut in the mountains, and the man shows us the fire pit for roasting the *piñas* of agave, his mule-powered grinding wheel, the barrels of fermenting mash, and the copper still that finalizes the process. Fernando loads a five-liter gas canister filled with the latest batch into his trunk. In Jalapa we stop by the bright-orange compound Fernando is building for his girlfriend and meet his young daughter. Both of Fernando's sons are grown and work with him on wind development. One of them tells us later that their father is spending a lot of time here in Jalapa. They are not entirely happy about the arrangement, but they accept it.

Between stops we talk for a while about wind power in Oaxaca. Fernando works for the Economy Ministry, so he says he views this mostly in terms of development. Oaxaca is the second-poorest state in Mexico, he reminds us, and the isthmus is one of the poorest parts of Oaxaca.

"What afflicts us today? That we need to develop our productive capacities in all respects. We need development. And that's the idea behind this wind energy. This is about developing the zone *through* electricity. It's very simple. The owners of the land, whether individuals or nucleos, have an important resource that can be developed to help generate other kinds of advancement."

Local resources plus translocal capital investment equals development, which in turn equals jobs, economic growth, infrastructural improvements, new educational opportunities, and improved health services. This is the calculus of neoliberal politics that undergirds private-public partnerships (PPP) as a model of social development throughout much of the world today.[1]

Fernando admits that "the *indígenas* (indigenous people) don't get as much for their land as they would in Spain or Germany or the United States. But the reason for that is not because the companies are cheating them but because there are no subsidies for this type of energy as there are in Europe. And we don't have the same kinds of excellent technical conditions for the grid that you do in the United States. So, overall, it's more expensive to develop the resource." He says that the going rental rate for uso de tierra is six thousand pesos per hectare,[2] and that is augmented if there are roads and turbines actually placed on the land.

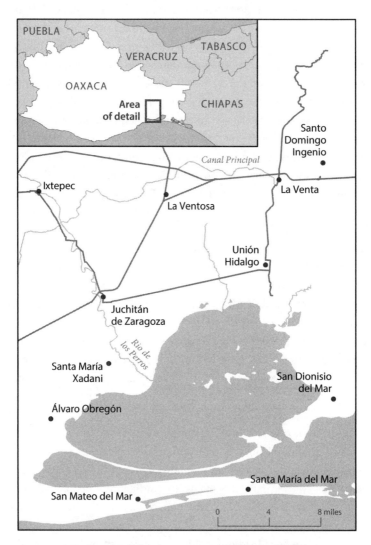

MAP 2.1. Core wind development zone, Isthmus of Tehuantepec

"So let's say a rancher leases twenty hectares to a wind park," he speculates. "That's 120,000 annually for the land, and if there's a turbine there too, it would bring him up to around 140,000 [$10,700] every year. That's a huge *asistencia* [help] because in this area, they can't really grow anything well because of the wind; it's just subsistence agriculture. This is win-win. There are no losers in this development. And it is also wonderful for the state. Now Oaxaca is becoming something new, the place in the country that is generating renewable energy."

But what about those in the isthmus who are criticizing the wind parks? we ask.

"There are people out there who say, 'Oh the Spanish, they are recolonizing us, just like five hundred years ago.' OK, you have to respect their position. Their sentiments are genuine. But it's not true what they are saying. The problem is that often there is no process of communication. One of the tragedies of this modern world is that you can just put your information, true or false, into an email or on the internet; you can make statements about things that are right or wrong, and others will read them and be influenced by lies. In this modern world, there's no longer enough of people sitting heart-to-heart like we are doing here and coming to understand things.

As we near Juchitán, Fernando begins pointing out the window, sketching an alternative modernity with his fingertips. He gestures toward green ranchland beyond the road that is destined to become wind parks. He names the park projects in development as we drive. He also mentions a fledging master's program in wind energy at the UNISTMO (the Universidad del Istmo) in Tehuantepec.

"It's a seed [*semilla*], something to sow. We want to develop more educational opportunities and jobs for educated people in the region. We'd like to create a city of knowledge [*ciudad de conocimiento*] here. Advancing education is important for both the companies and for our universities. Today we send our young people north to pick your strawberries and lettuce. In the future, they will go and run your wind parks."

It is a moving image of green energy building white-collar prosperity, engendered by a perfect marriage of private sector capital and expertise, the managerial talents of the government, and the abundance of istmeño resources and desires for a better life. Fernando's eyes water with possibility. Our conversations with Fernando over the next several years convince us that he is utterly sincere in his belief that his work to orchestrate wind power development is leaving a positive social legacy and a brighter future in the isthmus.

In the early afternoon we arrive, finally, in La Ventosa, some seventeen kilometers to the northeast of Juchitán. Fernando describes La Ventosa as a comunidad indigena, an indigenous community, poor, but one where people understand the benefits that this kind of development can bring.

La Ventosa sits toward the western side of the central wind corridor in the isthmus. The last federal census listed the population as forty-two hundred, but our La Ventosan friends tell us the real population is now closer to six thousand or even seven thousand. Many older La Ventosans are monolingual diidxazá (Zapotec) speakers, but the youngest generation is bilingual, verging on monolingual in Spanish: "Some never speak Zapotec, they only listen." La Ventosa sits at a fork where the highway to Matias Romero branches off the Carretera Panamerica. Farther down the highway to the east are the first two istmeño wind parks, La Venta I and La Venta II, both publicly financed parks operated by CFE. Meanwhile La Ventosa has become the epicenter of the new wind development model, privately financed autoabastecimiento (industrial self-supply) projects. The highway between La Ventosa and La Venta is famous for the fiercest winds in the isthmus, for gusts more than seventy miles per hour that topple massive semitrailer trucks almost casually. The local newspapers in the winter are filled with photos of their crumpled forms. There are places in the area, *zonas de turbulencias*, where no one would dare to put a wind turbine for fear it would be mangled by shearing.

The quality of wind is one of the reasons that autoabastecimiento wind development began in La Ventosa. But from a technical standpoint, there is excellent wind throughout the southern isthmus. Indeed, the wind in La Ventosa can be too strong for conventional wind turbines; they are often shut down when the wind reaches peak force to protect the equipment and the grid. La Ventosa represented an optimal location to pilot the new model for three other distinctive reasons.

First, in a region known for its fierce political divisions and conflicts, particularly concerning the disposition of land, for the past three decades La Ventosa has represented one of the most stable political strongholds of any party (in this case, the PRI) anywhere in the isthmus. Party leaders have also been openly friendly to transnational private investment since Miguel de la Madrid's presidency (1982–88), and three PRI governors of Oaxaca played important roles in shaping the process of wind power development in the isthmus.[3] La Ventosa was thus a promising terrain for investment that came complete with reliable political allies operating efficacious local patronage networks.

FIGURE 2.1. Fernando's favorite *mezcalero*

FIGURE 2.2. Trucks on the road between La Ventosa and La Venta

Second, although a considerable amount of the land around La Ventosa is bienes ejidales, it is an ejido that voted to parcel itself after the early-1990s land reforms under President Salinas. This meant that its constituents could now privately contract portions of ejidal land for their own benefit.[4] Together with an older group of politically organized, PRI-affiliated *pequeños propietarios*, this meant that a considerable amount of the land surrounding La Ventosa existed in something approximating a private-property regime, a regime more contested and simply uncertain elsewhere in the core wind zone.[5]

And third, La Ventosa is home to one of the most enthusiastic and long-standing proponents of wind development in the isthmus, don Porfirio Montero Fuentes, a local PRI and Evangelical Christian leader who is routinely described in the local news media as the town's cacique.[6] Montero rose from a relatively undistinguished family background to political prominence in La Ventosa largely as a result of his grassroots political campaign against the COCEI uprising in Juchitán.

According to local historian José Lopez de la Cruz, "It was the 1980s, and Porfirio talked a lot about how the reds would come and steal our land from us.[7] And then COCEI did actually try to occupy a large plot of land in La Ventosa, which belonged to an administrator from Juchitán. I remember Porfirio leading a group of armed men down the street, with guns, sticks, and machetes, and they eventually drove the COCEIistas off."

Somewhat ironically, Montero celebrated his victory by taking the majority of the administrator's land for himself and his allies. Now a significant landowner, Montero had become involved in organizing the cane workers of La Ventosa and was eventually made the Oaxacan state representative for the pequeños propietarios in Mexico City. This helped him gain access to the translocal networks of the PRI and to become involved in development planning for the region. Back in the isthmus, he cemented power in several ways: by accumulating more land,[8] seeking to control local politics, and building an Evangelical Christian organization, Cristianos en Movimiento, which was active in both local and state politics. In the late 1990s he also began to host La Gran Fiesta, the largest annual Evangelical meeting in the isthmus at his Rancho la Soledad south of La Ventosa.

It is not surprising, therefore, that the site of today's meeting concerning the Iberdrola project is being held on the veranda of Montero's church, Templo Evangelico Pentecostes La Nueva Jerusalen on the eastern edge of town. All around are half-constructed homes and cement poles marking out land plots. This area of La Ventosa is known as the Colonia Porfirio Montero. Fernando leaves us waiting with his driver while he assesses the situation,

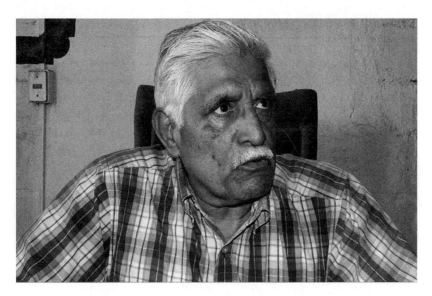

FIGURE 2.3. Porfirio Montero Fuentes

circling the assembled people, poking his head in here and there to listen, and shaking a few hands. Shortly thereafter, someone waves us over.

Montero, tall, broad shouldered, wearing a light plaid shirt and jeans, dominates the porch of the templo upon which the gathering is taking place. He is leaning springily over people's shoulders, joking, cajoling, channeling the moment. It is an important event because the company is distributing written estimates of what they intend to pay each of the dueños and ejidatarios for the impacts (afectationes) to their land.

Montero is preaching trust in the company. "And in the president of the Republic!" someone else shouts out.

Montero shrugs. All afectationes will be paid for in the proper amount, they have *his* word.

Two representatives from Iberdrola are sweating, not only because of the heat, and seem happy to let Montero run the show. Neither speaks a word of Zapotec. Fernando and an undersecretary for political relations hover in the background. In the end, their presence seems mostly symbolic. When Montero ends his speech with an invitation for everyone to sign the registry to show that they received their estimate, people begin to move toward the table.

Fernando looks visibly relieved and smiles. "It's an act of good faith," he tells us. "There will be time for them to review the details and raise questions, but the good faith is important. It's going to be a long process."

FIGURE 2.4. Wind turbines, La Ventosa, 2013

The undersecretary nods along saying, "The job of the government is to maintain equilibrium between the parties."

The deputy to the state congress for the region teaches us a few words of Zapotec and says confidently, waving his hands around, "This here is going to transform this town completely, when the wind parks come."

But the equilibrating mission of these agents of federal and state power, here as elsewhere, grinds against long-standing grievances and suspicions in the isthmus that government is incapable of doing anything other than feathering its own nest. Montero, for example, expressed his own doubts.

"We should be getting paid fairly," he said. "Look, we have no problem with these companies. It's the government that is making us fight them. At the last meeting we had with the landowners, and Iberdrola, the company, said, 'Why don't you ask your government for more support?' And we said to them, 'No one here wants anything to do with the government. They deceive us.' For ten years, I've been telling [the government] that they need to create a committee to regulate what is going on down here, and no one has listened to me. Why can't we be paid closer to the price that the government is paying [for electricity from the producers]? We're the ones producing the energy. . . . When you look at the generation that is going into their substation, we are only making about 1.5 percent. Listen, it's a dirty business."

Interestingly, Montero does not want to be paid by hectare of land but by megawatt, 100,000 pesos per megawatt. "This is what we are proposing to

FIGURE 2.5. Promotional image, La Gran Fiesta 2017

them."[9] He is moving from the logic of land to that of energy. But only half-way because he still views wind as a fixed material resource, much like land.

"It can be low, medium, or high, but the wind is always blowing," he said. Thus, he does not want to hear about variable payments but insists on a fixed one, "the same for all of us."[10] And he does not care whether that brings him a bad reputation "back in Spain."

He feels no one gives him the credit he deserves for the success. "I was the one who brought the companies here, I was the *iniciador* (initiator) of all of this because I'm someone who likes development. And I was the one who told them how the *aire* [air] is here."

But, again, he has no problem with the companies. "This is a marriage. It's the government that is going around beating on us. We don't want a divorce from these companies."

His latest plan is to convince the government to allow him and his allies to build a wind park that will sell directly to CFE as an independent power producer. It is not unlike what is being imagined in Ixtepec, just absent community ownership and the detailed social development mission. For Montero, it is about payments and business. But also about spirit. Since 2010, more and more wind turbines have been appearing in the promotional and archival imagery of La Gran Fiesta.

Already bleary, we face more interviews throughout the day. We tour the La Venta II park, realizing for the first time just how massive actual turbines are when we stand underneath them, watching their blades cut the sky, *thwup, thwup, thwup*. The government officials we speak with uniformly deny the existence of viable agriculture in the region. The parks are, at least in their minds, displacing nothing of value, conjuring prosperity from the air.

The only problem, the future mayor of Tehuantepec tells us, is the presence of *líderes* (political leaders) in the isthmus. "There are always líderes who want to start a movement against the government, against investment, basically against private capital. Here, those types abound." But he assures us that they are really just local political operatives who want a cut of something. Right now they are trying to negotiate a deal with the leader of the truck drivers' union in order to build the new substation in Ixtepec.

There is an interesting conversation with a Señor Velasquez, who used to work with Fernando in government and at his consulting firm and who has gone on to help Spanish firms like Iberdrola and Preneal manage relations with the propietarios. "The truth is that I have a good job, but a complicated one," he says. "I've seen the *empresas* [companies] arrive to promote things, and then propietarios not honoring their contracts. Often they don't read the contract, they want you to explain it to them, and then everything's fine. But you need more than goodwill because even though the companies may talk nicely about *la buena convivencia* [living well together], in the end, this is a signed contract. Maintaining the project is thus more about communication than permits and contracts. There are eighty-four propietarios in La Ventosa, no more than that. But getting those eighty-four to agree on anything is difficult. Some are from rival families, some belong to different [political] parties, and I have to convince them to *convivir* [coexist]."

Some hours later, we end up relaxing with Fernando in the crisply air-conditioned Café Santa Fé at the *crucero* (crossroads) in Juchitán. The Santa Fé is a curiously unassuming nexus of local aeolian politics; if its walls could talk, they would tell of thousands of wind-related deals that have been hashed out at its tables (see chapter 5).

Occupying a corner table, Fernando pulls out a laptop to show us a promotional video his office has just produced to educate people across the isthmus about the benefits that wind development will bring. The video opens with an old farmer and what we presume to be his grandchild staring out at a vast field of straw. Then there is a woman in traditional ikojts dress and a bare-chested fisherman casting a net down at the surf. Both stare at the camera confrontationally as if they are sick of their lot in life. Then a highway filled with traffic. Then a brightly lit modern building where three young people in casual dress smile confidently. Then a football field at night where children in bright uniforms wave flashlights in a circle imitating the rotation of a wind turbine. The entire video is a blur of crossfading images of modernity, nature, infrastructure, and youth, whirring like the turbines that appear throughout. At the end, a different farmer and grandchild point at a turbine, turn toward the camera, and smile, happy in their secured future. The course of development and modernity is relentless, uplifting, harmoniously bridging nature and culture, a perfect marriage of biopower and energopower, all enabled by the invisible (and undepicted) hand of capital.

Autoabastecimiento: A Love Story

How did the industrial development of wind power in the Isthmus of Tehuantepec begin? Why did autoabastecimiento become its dominant paradigm? These are questions that are easy to answer in broad conceptual brushstrokes of "capital," "development," and "energy." But they are surprisingly complicated and contingent when one begins to probe more deeply, not least because so many different actors and institutions played roles, sometimes coordinated, and sometimes worked at cross purposes. Certain narratives and memories resonate, others drown each other out, still others have been lost to the wind. And then there are questions that no one seems able or willing to answer.

Autoabastecimiento is a love story, that much seems clear. It is about capital, ever desirous of new opportunities for its own propagation, finding increasingly open doors and welcoming arms in the Mexican state. It is about that state's own desire to attract the enabling power of capital but to do so in such a way that would at the same time bring wealth, opportunity, health to its citizens and extend vital infrastructures of governmental power like grids and roads. Flirtation begot passion, which led, in turn, as Porfirio Montero might have it, to marriage. And marriage is an exchange relationship unto itself, perhaps not exactly what either lover fantasized, yet (they hope and tell themselves) an enduring re/productive partnership.

The romance unfolded largely in the 1990s and 2000s. But some key elements predated it. In isthmus communities in the core wind zone, some recall that *extranjeros* (foreigners), government officials, and researchers began to appear periodically in the 1970s to talk to them about developing wind parks. The 1970s were, it is widely known, a critical, paradoxical energopolitical moment across the world.[11] The formation of OPEC and the oil shocks rapidly destabilized the carbon-political order that had thrived in the middle of the twentieth century, exposing the vulnerability of the exercise of political power and biopolitics upon the basis of secure energopolitical infrastructures and flows. States anxiously sought to reestablish the image of political autonomy and sovereignty—from other states but also, importantly, from fossil fuels—in part through new projects of fossil fuel development and in part through greater experimentation with energy conservation and investment in nuclear and renewable energy forms.

Mexico was, by and large, a beneficiary of the oil shocks. As a nonaligned country that was friendly to the global North and, more importantly, as a non-OPEC member with massive oil reserves, Mexico found its oil heavily in demand after the early 1970s, which helped spur a quadrupling of Pemex's production capacity by the end of the decade. Moreover, all profits were retained by the Mexican state because of the 1938 nationalization of oil. A Mexican petrostate was by no means conjured from thin air in the 1970s, but it was the period in which Mexico transitioned from a net importer to an exporter of oil. Oil revenues quickly became essential for the fluid operation of Mexican statecraft.[12] With so much attention focused on oil in the 1970s, it perhaps makes sense that wind power remained at best in the peripheral vision of the Mexican state.

Yet there was decisive activity all the same. In the early 1970s, Mexico was a leader in the G-77 bloc of developing nations in the United Nations and was instrumental in the 1975 UNIDO (United Nations Industrial Development Organisation) Lima Declaration. The declaration essentially argued that independent industrial growth was the priority for the developing world given the "persistent and marked tensions to which the present international economic situation is subjected."[13] Among other specific recommendations, Lima advocated that developing countries build their electricity industries, support rural industries, and stimulate processes of applied scientific research.[14] A few years later, a research team led by Enrique Caldera Muñoz of the Instituto de Investigaciones Eléctricas (IIE) began researching wind potential in the isthmus and published a preliminary study focused on La Ventosa in 1980.[15] From 1979 to 1983, Caldera's team also worked collaboratively with wind energy researchers from several other countries under the auspices of the Organización Latinoamericana de Energía (OLADE) to produce a first provisional atlas of technical wind capacity across Latin America and the Caribbean.[16]

Given surging oil revenue, the context of Lima, and the fact that Caldera's team was working in a research unit of the Mexican Ministry of Energy

(SENER), it seems highly likely that any serious discussion of a program of wind development in the late 1970s or early 1980s would have focused on a public-financing model. The public model would have brought the program under the auspices of CFE—since the nationalization of electricity in 1960, CFE was the only entity legally capable of producing, transmitting, and selling electricity in Mexico. But our interviews and archival research yielded no sign that either CFE or SENER had formulated any concrete proposal for wind parks during this period beyond acknowledging that Oaxaca had exceptional resources. A rare study of the early years of Mexican wind energy laments the missed opportunities of the 1980s: "At that time, very few people believed that the wind could become a source of energy for generating electricity at any significant level."[17]

The window of opportunity for a public model of wind power development also passed quickly. The global oil glut of the early 1980s generated widespread social and economic crises in Mexico with sagging Pemex revenues, high unemployment, and 100 percent annual inflation for most of the decade.[18] There was very little new public infrastructural investment of any kind during the presidency of Miguel de la Madrid (1982–88),[19] who also led what has subsequently become known as the "neoliberal turn" of the PRI party and the Mexican state toward reducing public spending, privatizing public industries, opening markets, and attracting international capital.[20] Caldera struggled to find funding for an anemometric study of wind capacity but, with the help of CONACYT,[21] he was eventually able to set up five measurement stations in the isthmus between 1984 and 1986. Their results were quite promising, especially at the La Venta and La Ventosa stations. The average annual wind speed was 9.3 meters per second, with top speeds in the winter around 30 meters per second.[22] But the data had no signal in the political turmoil encompassing Mexican statecraft for the next several years until the neoliberal turn had established itself more securely.

Exposing Mexico's lucrative energy sector to international capital has long been an objective of neoliberal reformers in the Mexican government.[23] But it is an objective that has been hampered by the national sentiments associated with fossil fuel resources, so much so that even after nearly thirty years of Mexican neoliberalism, genuine profit sharing from Pemex's fields remains highly politically contentious today.[24] However, the Mexican electricity sector proved comparatively easier to open to foreign capital. Presidents Carlos Salinas de Gortari (1988–94) and Ernesto Zedillo (1994–2000) pushed hard, and with some success, to open the sector to private interests, at least in terms of electricity generation.[25]

Salinas founded the Comisión Reguladora de Energía (CRE), a new energy regulatory body to, among other things, promote and manage private-public partnerships in electricity generation. These partnerships came in a number of types, but the two most salient were (1) independent power production (PIE), where independent electricity producers sold directly to CFE, and (2) "power generation for self-sufficiency," or autoabastecimiento, in which large electricity users (like industrial concerns and even Pemex) could enter into long-term purchase agreements with independent electricity producers for fixed electricity prices while sometimes owning a stake in the production company as well (with CFE acting solely as a medium for the transmission of electricity across its grid).[26] In its twenty years of operation, CRE has, by its own accounting, been able to privatize approximately one-third of electricity generation in Mexico (according to a January 2014 report from CRE, CFE accounted for 62.7 percent of generation, PIE for 21 percent, and autoabastecimiento for 8.7 percent).[27] Privatizing electricity generation allowed the federal state to extend infrastructure (see chapter 1) and expand capacity while circumventing the cash-strapped and often-recalcitrant bureaucracy of the public utility (see chapter 4). But love, unlike desire, is properly a two-way street. Bankers in Mexico City told us that because of CFE's very high electricity tariffs, a private investor in the wind sector, depending on when it entered a project and for how long it invested, could safely expect a 10 to 20 percent return on investment, a very lucrative, low-risk arrangement.

As the main love story took shape in the 1990s, other, more idiosyncratic partners emerged as well. For example, there was the colorful Carlos Gottfried Joy, who claims to be one of the "fathers of Mexican wind energy." Gottfried, an engineer and environmentalist, founded a small wind turbine firm in the late 1970s that was based in Mexico City but oriented toward export to the growing US market for wind power. The firm failed, in part because of the reliability of his turbines, but Gottfried claims that the experience inspired him to bring wind power to Mexico.[28] Later a millionaire after taking over the electrical equipment company cofounded by his father, Gottfried developed the idea of building a Mexican market for the turbine generators and other electrical elements manufactured by his company and its partners. With co-investment from Enron Wind, Gottfried founded the first Mexican wind power development company, Fuerza Eólica, in the early 1990s and won a tender from CFE to build a 1.5 megawatt pilot municipal wind park on the Cerro de la Virgén in Zacatecas. The project collapsed "in a practically Kafka-esque way" according to a contemporary newspaper account,[29] with much finger pointing over debts and a very public 1997 court case in which Enron was caught trying to manipulate the case's outcome via the Clinton administration. Eventually cleared of charges, Gottfried returned, unbowed, in 1998 to receive the first two CRE permits for autoabastecimiento wind parks, including the first for Oaxaca.[30]

Meanwhile, collaborating with the Danish firm Vestas, CFE was eventually able to build its 1.5 megawatt pilot park, La Venta I, which went into operation in 1994. The data from the park seems to have impressed even the skeptical CFE. By comparing energy production data from La Venta I with 1,600 other identical turbines placed elsewhere in the world, CFE and Vestas concluded that in its first year of operation, the park operated with a plant capacity of 51.7 percent,[31] a figure surpassed by only one installation in New Zealand. The cost of electricity compared favorably with thermoelectric installations elsewhere in Mexico ($0.04 per kilowatt-hour). The outstanding technical data began to draw attention to the isthmus, whetting appetites for investment.

Although perhaps not the most effective developer of wind parks, Gottfried appears to have been skilled at assembling a lobby of industrial interests to sponsor autoabastecimiento projects from which they would in turn receive guaranteed electricity supply. The late 1990s was a period during which the political and industrial interests behind Mexican wind power began to consolidate, with all eyes on the prize of harnessing the winds of Oaxaca. Diódoro Carrasco, the governor of Oaxaca at the time, told us that he did not recall anyone mentioning wind parks to him when his term began (1992) but that by the end of his term (1998), "it was becoming a topic."

In 2000, working in the government of Carrasco's successor, José Murat, Fernando Mimiaga began actively promoting the development of the wind corridor.[32] In addition to his governmental work, Fernando and his sons founded an NGO, the Fundación para el Desarollo del Corredor Eólico del Istmo y para las Energías Renovables, which acted as both an advocacy and consulting entity.[33] The fundación helped the government coorganize seven international colloquia beginning in 2000 for the "public promotion of investment opportunities in wind energy."[34] Fernando claims that he brought fourteen foreign companies to the isthmus in 2000–2001 alone to explore investment opportunities.

Forces rapidly began to assemble to remove the remaining barriers to privately financed wind power. President Vicente Fox's effort in 2001 to redefine electricity from a public service to a commercial service failed. But CRE unveiled a model contract the same year for connecting independent renewable energy producers to the grid. In 2002, SENER prompted CFE to begin planning a second public park with technical assistance from a private energy developer (Iberdrola) and financial assistance from the UN's Clean Development Mechanism. Not long thereafter, in August 2004, at the tony Pacific resort of Huatulco, the lobbying organization Asociación Mexicana de Energía Eólica (AMDEE) was founded, with Carlos Gottfried unanimously elected as its first president. Both AMDEE and the Fundación take credit for pushing the state government of Oaxaca to undertake a program to "regularize land tenancy" in the isthmus in order to reduce investor risk and for designing a new grid and substation model that would allow electricity from the wind parks to be evacuated from the isthmus. They collaborated closely on these projects with a number of state and federal agencies (CFE, SENER, SEMARNAT, SEGEGO) and with local advocates like Montero, who also describes himself as the person who convinced the state government to give property owners a more secure title to their land during the governorship of Ulises Ruiz (2004–10).[35] These processes did not unfold without controversy. Both AMDEE and the Fundación have also been accused of cartelism and corruption by opponents of private sector wind development,[36] specifically for parceling out plots of land in the isthmus to developers with little or misleading consultation with residents.

In 2003, the US National Renewable Energy Laboratory (NREL), working with the Oaxacan state government and SENER, published a comprehensive wind atlas of Oaxaca,[37] which confirmed and extended the earlier findings from La Ventosa. The isthmus was now internationally recognized as having some of the best technical wind resources anywhere in the world. Fernando told us that the NREL report helped greatly to "strengthen the political will to break the barriers impeding development." The last piece of the puzzle was Calderón's 2008 Ley para el Aprovechamiento de Energías Renovables y el Financiamiento de la Transición Energética (LAERFTE), which not only forced SENER to set binding targets for renewable energy development but also gave CRE increased authority to oversee the development of the renewable energy sector free from the bureaucracy of CFE.

The wedding preparations were completed. What transpired between 2009 and 2012—when we returned to Mexico for the main phase of our field research—was something of a honeymoon period. More than thirteen wind parks came on grid in the isthmus during those three years with more than 1.2 gigawatts of new capacity, ten of them autoabastecimiento projects.

But why autoabastecimiento and not PIE, which some developers told us was much easier to manage (since there was only one purchaser, CFE, to deal with instead of several)? Because Mexican industrialists and banks were interested in having a lucrative ownership stake in wind park projects and because various large corporations in the mid-2000s wanted to be able to advertise that they were investing in renewable energy at a time of mounting public concern about climate change. Those desires were further fueled by the promise of credits for certified emissions reductions (CERS).

As one banker put it, "In the good old days before the market collapsed and they became essentially worthless, CERS could actually amount to perhaps 7 percent of the total investment," more than $30 million for a 200 megawatt park. Thus the aspiration to protect the bios of planetary life was intertwined with the logic of profit and also with a more purely energopolitical interest in ensuring steady energy supply.

A friend of ours, a Mexico City–based journalist with many years of experience covering the energy sector, explained, "A lot of people here still remember the old days when there wasn't enough electricity and CFE would sometimes just shut a factory down without warning. . . . And the industrial electricity prices here are still 30–40 percent higher than they would be in the United States. That makes the self-supply model attractive. It reduces risk in terms of both price and supply."

The Iberdrola project in La Ventosa was, as noted above, the first of the new privately financed wind parks to be consummated. Like many others, it was many years in the making. At the time it received its CRE permit in 2002, Iberdrola was not named as a partner. The application was made by a group called Parques Ecológicos de México (PEM) consisting of industrial concerns (including Pemex, Cementos Apasco, Hylsa, and Volkswagen) seeking to self-supply with clean electricity. In November 2004 it was announced that Iberdrola, at the time the second-largest electrical utility in Spain, had purchased the project, its first wind park outside of Spain, and would make PEM its Mexican subsidiary. The newspaper *La Jornada* wrote a needling editorial describing the seven-million-peso payment Iberdrola made for the park as a pittance, suggesting that this was more evidence of CRE's mission to divide "*las rebanadas del pastel*" (the slices of the cake) of the Mexican energy sector among hungry international suitors.[38]

That hunger endured when we returned to the isthmus again in the summer of 2012. It was now clearer than ever that wind power development in southern Mexico was a lucrative enterprise. But it was also becoming a more hotly contested one. Conflicts were multiplying to the south of Juchitán about the planned use of communal land in the Mareña Renovables and Gas Fenosa park projects.[39] And even in La Ventosa, a town in which so much had indeed changed, with turbines visible in all directions, opinions differed considerably as to whether what was happening was helping la comunidad or only a handful of landowners.

Guidxhi Ra Riale Bii

Local history dates the founding of La Ventosa to November 23, 1846. The land the town now occupies was still virgin forest with just two unpaved paths cutting through the area. Along one of these paths came Felipe López Lucho and Marcela Martínez de Córdoba (he born in Spain, she in Mexico) looking for new grazing grounds for their cattle. They settled by a tributary river known as Lagatero Viejo north of the current town center, and "not being deterred by either the strong winds that broke branches nor the persistent howling of the coyotes,"[40] the young couple raised nine children and hundreds of cattle on their ranch. It was thirty-five years before the next family moved to the region, and then over the following thirty years, a small nucleus of families slowly gathered, creating "a poliform culture constituted of Zapotecs from Juchitán, Ixtepec, and El Espinal, Zoques from Chimalapas, as well a plurality of persons originating in Tabasco, Chiapas, and Veracruz, . . . who adapted themselves to the dominant Zapotec culture and customs of the region."[41] They ranched, farmed a local variety of maize called *zapalote chico*, and hunted iguana, armadillo, deer, quail, and rabbit. They gathered cascalote seeds and made charcoal to sell in Ixtepec. The Zapotec name for La Ventosa, proudly emblazoned on many of its taxis and motos today, is Guidxhi Ra Riale Bii, "the place where the wind is born."

La Ventosa seems to have remained a loose-knit settlement until the first La Ventosan *vela* celebrations took place in the first decade of the twentieth century, marking its consecration as a true istmeño pueblo.[42] Lacking a church, La Ventosans celebrated their velas in Juchitán instead until 1926, a period in which the ties between La Ventosa and Juchitán were intimate but also contentious. The Mexican Revolution was highly polarizing when it came to the isthmus, as elsewhere in Mexico, leading to a general opposition between the elites of Juchitán and their networks who supported the Constitutionalist cause of President Ventusiano Carranza (known locally as the *partido rojo*, red party) and the lower classes of Juchitán and the surrounding region (known as the *partido verde*, green party), leading to a number of uprisings and armed conflicts between 1910 and 1920. When Oaxacan governor Juárez Mata sent troops to occupy Juchitán in 1911 to quell the

FIGURE 2.6. The place where the wind is born

Chegomista uprising, the resulting violence and instability sent many families fleeing to resettle in surrounding villages like La Ventosa. La Ventosans remember the governor sending soldiers to burn their homes too, claiming that the town was *puros ladrones* (nothing but thieves). In 1919–20, when future general Heliodoro Charis organized the verde San Vicente uprising in the marshes south of Juchitán, he attracted La Ventosans to the cause of unseating the rojos and their Carrancista allies from power in the region.

When Charis occupied the town hall in Juchitán (the same day that Carranza left office in Mexico City), he began the process of uniting the rival political factions under his charismatic authority.[43] The caudillismo and "Zapotec sovereignty" that Charis asserted over the region throughout the next four decades were certainly felt in La Ventosa,[44] as were the major infrastructural projects that he brought to the region as mayor of Juchitán and as federal deputy: the building of the highways; the establishment of wells, running water, and a sewage system in the 1950s; and not least, the gradual electrification of the town. The charcoal trade for which La Ventosa was well known throughout the isthmus diminished as other kinds of fuel became available. And new roads meant migrant labor to the sugarcane fields of

Veracruz or the cotton fields in Chiapas. The redistribution of national oil wealth in the 1970s brought new federal irrigation projects to the region and a sugar refinery in nearby El Espinal. Laborers returned to La Ventosa and began growing rice and sugarcane.

Wealth flowed. "Think about this," one resident recalled. "There were thirty-three cantinas in those days for a town of maybe two thousand people. And so every weekend the town was filled with drunks."

Porfirio Montero and his parents converted to Evangelical Christianity during this era, and Montero began to build his wealth and influence among the sugarcane growers. Adobe and palapa houses rapidly disappeared and were replaced by cinderblock and concrete structures. But the boom was not long lasting. Rice cultivation did not fare well and ceased. The sugarcane was hit by a major rot that wiped out nearly all the local fields. "In the 1990s," a resident recalled, "people went back to ranching again, and from that, cheese making. There was nothing else for them." The fierce winds made it challenging to grow a marketable crop other than sorghum. Like so many other pueblos in the isthmus, La Ventosa experienced short periods of tantalizing prosperity interrupted by long periods of struggle for subsistence.

Another kind of struggle that defined life in La Ventosa was the struggle for autonomy from Juchitán. Charis's influence cast a long shadow; he contrived to make La Ventosa a township of Juchitán in the early 1950s, the legality of which is still disputed by some. Moreover, Charis kept political parties out of La Ventosa since he viewed the parties, and particularly the PRI, as threats to his personal authority.[45] Even the federal government was loath to intervene in Charis's domain.

That changed almost immediately after Charis's death with the plan (or imposition, some would say) of the Don Benito Juárez García Dam near Jalapa del Marqués, a town northwest of Juchitán. The dam was a federal development project whose main ambition was to create an irrigation system designed to allow tens of thousands of hectares of land in the isthmus to receive adequate water for more intensive and profitable forms of agriculture. In the years before the dam was built, land speculation swept across the isthmus, encouraging the wealthier smallholder peasants (including descendants of the pre-Revolutionary caciques) to accumulate land. Speculation had various effects including consolidating estates, swelling the ranks of landless peasants, and creating sharper class distinctions between landed and landless peasantry.[46] In particular, the political leaders of Juchitán exploited the new conditions by assembling small plots into massive private landholdings for themselves. The federal government recognized the potential for serious political unrest and intervened with two federal decrees designed to limit speculation in the traditional bienes comunales of Juchitán. The first decree, in 1964, held that the vast (64,000 hectare, 247 square mile) bienes comunales were now ejidal owing to the inclusion of some 28,000 hectares in the irrigation zone. However, the infelicitous wording of the decree also seemed to deny any possibility of smallholder property, which caused the smallholders, in turn, to organize politically against the state. The second decree, in 1966, reinstated smallholder titles in the form of issuing some 25,000 permits to pequeña propiedad on the ejidal land. But the decree left the relationship between possession and ownership of land rather murky.

Shortly thereafter, COCEI, a political movement with deep roots in the landless peasantry, formed using land occupation as its chief political tactic, seizing what it viewed as inalienable communal lands back from the landed peasant class.[47] Land tenancy suddenly became a burning political issue in the isthmus, and absent Charis's caudillismo, old lines of political conflict shone brightly again. The federal and state governments faced off against the istmeños, the elites of Juchitán against the elites of its townships, the campesin@s and fisherfolk against the commercial and political class.[48] In La Ventosa, a town born with no ancient núcleo agrario but instead as a loose-knit community of homesteaders, the PRI were able to enlist many local landholders, particularly the influential founding families, in defense of the logic of smallholder private property. But Juchitán itself was soon riven by fierce conflicts between the PRI and the Juchiteco elite on the one hand and on the other those who wished to recuperate, through violent confrontation and occupation if necessary, communal land lost to privatization. From the early 1970s to the early 1980s, the PRI-COCEI battles became legendary and, some would argue, nationally significant, as Juchitán became the first city in Mexico to vote the PRI out of office, reinstantiating a long-standing reputation for ungovernable violent indigeneity (see chapter 5).

In La Ventosa, the old political distinctions between rojo and verde were reactivated. Montero invoked the language of international communist conspiracy to discredit the "red" COCEIistas and mobilize his smallholder allies. What also resonated in La Ventosa was the idea of throwing off the yoke of Juchiteco power in any form. A landowner who considered himself politically unaffiliated said, "It's always a clash [choque] with Juchitán. They are a disaster, a disaster. The mayors there don't do anything for us, but they still try to influence the citizens here."

In 2012 we wander the streets of La Ventosa admiring new brightly colored homes and cars. The road next to the house we have rented is being

FIGURE 2.7. New homes, La Ventosa

paved courtesy of Iberdrola and the agencia government. A second wind park, Electricidad del Valle de México (ninety megawatts), operated by the French firm EDF, has started production partly on La Ventosan land, channeling more money to local propietarios. But the politics of land and value echo across the decades. The PRIistas from La Ventosa and Juchitán are now battling over the money generated by the wind parks, specifically, over the administrative fees that wind developers pay for the change of land use, which go directly to the municipio in Juchitán rather than to the agencia.

The comisariado of La Ventosa's ejido,[49] Pedro Castillo, acidly reckoned with the injustice of that situation. "Thanks to this foreign investment, we are breathing, we are flowering here," he said. "But only little by little, *con lucha* [with struggle]. All the impact is happening here in La Ventosa. Shouldn't all the economic benefits be coming here as well? Well, because of the law, the change of land use can only be decreed by the mayor of Juchitán, Dr. Daniel Gurrión. But day after day, night after night, for twenty-five years, it is we who will be hearing the sound of the turbines droning. Not Gurrión, who lives seventeen kilometers away. I don't find that acceptable. We should be getting 100 percent of the money that is now being paid to Juchitán. Our agente municipal [Hageo Montero], was able to cut a deal finally to get 50 percent for La Ventosa of the probably eight or nine million pesos the companies are paying for permission." Prosperity beckons, yes, but still with so many questions, doubts, and problems encircling it.

Pedro is animated, his forearms chopping up and down, "I told Geo [Montero], I said, 'Fine, cut that deal with him, tell Gurrión no problem.' So he'll sign the document. And then when [Montero] has the check in his hands, [we'll] blockade the highways. And demand 100 percent. Doesn't matter who pays it. The state government can pay it. Or we are staying here, staying right here."

Pedro sinks back on his couch—comfortable, modern, leather—with his arms straight out, palms down. "That's what *I* would do if I were a militant PRIista. The government doesn't understand words; it only understands fists. All these politicians, they live from deception of the people."

"It's 8.5 on a Scale from One to Ten"

A kind and generous friend, Rusvel Rasgado, helps us to get situated in La Ventosa when we arrive back in the isthmus in 2012. We wish to talk in more depth to the people who have rented land for wind development. And we also want to complete a full community opinion survey of all households on how they view the impact of the parks. Thin, handsome, and charming, Rusvel is a local journalist who covered the wind parks from early on. Increasingly he has been able to augment his writing with broadcast work as a local reporter for the Mexican national broadcaster, Televisa. He was born and lives in La Ventosa together with his wife and children. Among his many cousins is agente municipal Hageo Montero, son of Porfirio. Rusvel makes a call, and we are on our way.

The agencia building is modest but centrally located. A few police officers are standing around outside. We are ushered right upstairs to Hageo's office. It is strikingly bare, containing only a desk, three chairs, and a cabinet sagging under stacks of yellowing documents. On his desk is a worn paperback copy of the federal penal code and, under it, a Bible.

Hageo is a quieter, gentler presence than his father. He tells us that like everything else in the world, for example, building a house, the wind parks have good and bad impacts. "There has been a lot of change in the community. I now see young people who are finishing their studies in Juchitán as electrical engineers who are working for the companies now in very important posts.

They have taken courses in Spain, in Tijuana. They are building houses that are better than their parents'. They are making the most of this."

Fernando Mimiaga's dream is coming true, at least for some. And Hageo sees general social benefits as well. "At a global level, we're helping to generate clean energy." And locally the idea is to do one major project (*obra*) per year as a partnership between the agencia and the company. This year they are a paving a street. Next year they will open a new cultural center, "where there will be classes for children to learn to play guitar, to learn to draw, where they can hear lectures."

Hageo remembers the days when horses rather than cars were the dominant form of transportation in La Ventosa. But he says the land has always been *inservible* (useless) because of the wind. They could only farm in the months when it abated, "really only May, June, and July." Children could see no future in agriculture and ranching—"It's hard work; they want to be doctors now or engineers"—so they were leaving the region, and no one was caring for the land. But now, he says, they are returning because of "el eólico" (as he puts it) and even fighting with each other and their parents over inheritances. Hageo returns to these fights (*pleitos*) more than once in our conversation; these rifts within families and across generations seem to trouble him. He also finds the Spaniards somewhat *enclavado* (closed off). "They don't like to put our people in the best jobs, they keep those for themselves."

But little by little, they are trying to overcome that; he has a nephew now who is highly placed at a park. And meanwhile the construction phase of the parks brings a lot of work to the community, particularly to those who own trucks and work in the construction union. So they are planning to build two new parks with the Spanish firm Gamesa, Dos Arbolitos (seventy megawatts) and El Retiro (seventy-four megawatts).[50] Hageo is not sure about the environmental impact of these parks—he wonders at the quantity of concrete being poured into the soil. But he grins a little when he talks about his cattle. "I think they like the turbines for the shade they give. I see them clustered out there all the time, moving with the shadows as the sun comes across the sky. The noise doesn't even seem to bother them. They're eating well."

Our interviews with other propietarios turned up a range of thoughtful responses to the parks. All spoke warmly of the new recursos (resources) that the parks brought, of the potential for economic and social development. Logics of kin and land had spurred interesting contractual innovations, some of which seemed to be unique to the isthmus. We discovered, for example, that because the rental contracts were negotiated individually with the

FIGURE 2.8. José Lopez de la Cruz

smallholders, some of them had been able to make special deals with the companies, including long-term employment for their children or grand-children, good white-collar jobs that would keep them close to home. We also discovered that istmeños had innovated a new kind of land payment for *derecho del viento* (right of wind), a kind of easement for the wind that passed over one owner's land en route to a turbine positioned on another owner's land.[51] These kinds of extra payments, wind developers and bankers in Mexico City would later tell us, helped to keep the peace in a contentious political environment.

José Lopez de la Cruz is another of Rusvel's cousins and the town historian. He created La Ventosa's only documentary film and is revered as an authority on local culture. He screened the film for us in a garage he had assembled for his bright-red Volkswagen Bug from materials recycled from the wind parks, including foam transport spacers used for moving turbines by truck and planks of wood emblazoned brightly with a blue Vestas logo. He explained to us that when the parks were first proposed, there was resistance and conflict.

"The politicians from here, they wanted money. So they played with people, deceived some of them, told them that the wind project was not

FIGURE 2.9. José Luis Montãno.

viable. Then, when the first private wind park was announced, people came from Matias Romero and marched here against it,[52] but these people had no relationship to the lands of the isthmus."

Overall, he assured us, La Ventosans had taken a long-term view of the project's benefits. "In truth, we here were always in favor of the project because we knew that this was something that would benefit everyone directly or indirectly. First, obviously, it would benefit those who had land. That's logical. But then it would come to those who had businesses, and then to those who sell food, and then to those who sell food at night, all of them."

And migration had ceased, indeed reversed. "The only bad thing I can see about the parks is that the housing rents here have shot up. It used to be so cheap to rent a home."

Other propietarios highlighted various social issues that they felt kept the development process from being a pure boon to the community. We interviewed José Luis Montaño on his veranda as late afternoon turned toward dusk, speaking until he had to reach up and switch on the incandescent bulb hanging over us. In his late fifties, Montaño had three plots of land (*predios*) and described the coming of the parks as a *bendición* (blessing), not least because the companies were finally able to help them secure land titles after a long history of conflict over land tenure. He walked us through that history in detail but described himself as a simple campesino, someone who grew up knowing only the farming and ranching of his father and grandfather. It was hard, tiring work. And then the wind parks came. He talked softly, often searching the sky.

"The advantages? Well, I suppose one of them is that the land rental covers one's needs. You don't need to work the land. To wake up in the very early morning [*madrugada*] to care for the crops. There's less risk, if the crops fail. . . . They've improved the roads. The town has changed a lot."

How? we asked.

"The streets have been paved . . . ," he trailed off for a moment. "Health care, well, I won't lie to you, that that hasn't improved, and it should be a priority. The majority of people here don't have it. We need a real hospital here like in Juchitán, not just a health center. Too many people have to travel for care or go without care. So I think that health cannot wait. It cannot wait!" Whether the government or company provided the health care mattered less to him. "Our health is the most important thing. It's the first gift God gives us."

Montaño also noted that the wind parks have led to stratification even among landholders. "Some of the ones with more turbines, five or seven, might have an access point from the road for the maintenance of the turbines. And they threaten to close that off unless the company pays them a certain amount. But not everyone has that ability to close things off, not everyone can pressure the company in that way. It's not just."

Still, in terms of La Ventosa's overall experience with the wind parks, he said as he grinned in the twilight, "It's 8.5 on a scale of one to ten."

Cain López Toledo is likely the most critical voice we hear among the landholders. He says bluntly that he feels life was better before, when people worked together in the *campo* (countryside). "So much money divides people, and it's divided our people." The money only goes to people like him who have land. Others may get a little out of the construction phase or into their business, but he feels that is very little.

And the easy money from the rents has made the landowners *floja* (lazy). "There are people who just live on the rents now. For me, I live from my cattle. I have my ranch. I grew up in the campo, and I like my life in the campo. . . . Now, well, our way of life has already changed into something that is different."

He waves off those who complain about the negative effects of turbines on his land, and he has no problem working with the company. The problem is the rising inequality among people. As we are getting up to leave, he tells us how much he enjoys the local corn, the zapalote chico—it grows fast when the wind is low; its stalks are short, but the corn is so sweet and white. But as fewer and fewer people farm in the region, the price of the corn has been rising, and it is harder to get the local corn. "And for the poor people who live here, the people who have nothing, it's getting harder and harder to live here. With the corn costing six, seven, eight pesos a liter, maybe even more, for this corn, our native corn."

Everyone Comes Apart

In March 2013, as we begin to walk the streets of La Ventosa, clipboards in hand, knocking on the doors of houses both prosperous and humble, speaking with people both animated and disinterested by our questions, a more complex landscape of local opinion begins to take shape. As part of a more comprehensive survey project, Cymene and I personally visit eighty households, representing 370 people, or about 5 percent of the estimated population of the town. Whereas with the landowners we had spoken mostly to older men, here we have a chance to listen to younger men, the majority identifying as *obreros* (laborers) and campesinos, as well as a great many women: vendors, store owners, cooks, and *amas de casa* (housewives). It quickly becomes clear that opinion is highly divided about the impact of the parks now four years after Iberdrola's Parques Ecológicos de México began operating.

We ask interviewees to recall how they felt when they first heard about the projects, whether they thought it would be something good for La Ventosa and for their families specifically.

In the beginning there was hope, expectation. Several, 38.8 percent, say they remember feeling that the wind parks would be good for the town, and 43.8 percent say they expected a positive benefit for their family in particular. Fewer recall anticipating negative impacts (33.8 percent and 26.3 percent respectively). It is striking that a significant percentage of respondents (30 percent) felt that the development would be *sin importancia* (of no consequence) for their families, and all of these say it was because they lacked land. Those who remember hope associated with the parks tell us they hoped for progress and cheaper electricity but, above all, for new recursos and sources of trabajo (work) in a town that offered few opportunities. Those who expected nothing good from the parks cite a wider range of concerns, ranging from negative health and environmental impacts to the expectation of more fights over land and, of course, more highway blockades.

When we ask whether the parks brought the benefits they had hoped for four years ago, only 20 percent say yes, 60 percent say no, and the remaining 20 percent say *en parte* (partly) or express no opinion. Of those we interview, 26.3 percent say their families received direct benefits from the wind industry in the form of a land contract or work. Of those directly benefiting, just over half (52.3 percent) rate their experience with the companies as a "good" or "excellent" experience. Another 28.6 percent rate their contact as "so-so" and 19 percent as "bad" or "very bad."

When we ask whether the companies and the government provided enough by way of obras for the good of the town, we encounter strongly negative opinions. Only 25 percent say that the companies are doing enough by providing paving. "It's not nothing with this wind that whips the dust up into your eyes." But 66.3 percent opine that the companies should be doing more; the obras are a *misería* (pittance), or things that had been promised like a new school never appeared. The paving was done on the cheap, without drainage, so that when it rains, there is more flooding than before.

Still, the worst criticism is reserved for the government. Only 13.8 percent say the government has done enough for the community whereas 83.8 percent say it has not. "The government? They never do anything for us. Here *se falta mucho*" (we are lacking a great deal).

But the percentages only trace the surface of the ambivalence regarding the parks here in the birthplace and heart of autoabastecimiento wind development. Why does dissatisfaction run so deeply? We are fortunate that several families invite us out of the branch-bending wind and into their homes, to sit with them and talk in more depth. Some are ardent supporters, others ardent critics, and their thoughts put flesh on the numbers.

The supporters tend, by and large, to be families that own land that was contracted or who received work from the parks in the construction phase or, more rarely, in the operations phase. They do not tend to speak at great length of the contribution of wind energy to the battle against climate change, but they do talk about modernity and progress, about how children can now aspire to jobs as engineers. The paving of roads was welcome, and they expect more obras to come, just as Hageo promised. The parks are a blessing, a change, yes, but a change for the better, as one can see in the quality and scale of the new houses being erected in the community, in the abundance of new vehicles, in the national importance of the town as the center of a new industry.

Those who are critical of the impacts on the community tend not to be critical of the wind parks themselves but rather of how the income from the parks is not being redistributed more widely. Countless individuals express some variation of the sentiment, "There are no benefits for people like us."

Some say they believed they would receive something: "At least that the electricity would be cheaper, but if anything, it's become more expensive."

Others say they never had any such expectation: "No one really had much hope who didn't have *terreno* (land) in the first place."

All agree that the wind parks are contributing to greater inequality between the dueños and everyone else. Few blame the companies directly for this situation, although we do sometimes hear the language of the *segunda conquista* (second conquest of Mexico by Spain).

More blame the government (local and federal) for allowing it to happen through their neglect and abandonment of the town. "Those who govern are the only ones who benefit. I've heard that only 1.5 percent of the profits goes to the campesino."

But the agents most often singled out as the cause of the lost opportunities are the political líderes, principally Montero and the PRIistas, but also Mariano Santana of COCEI, exercising influence from afar in Juchitán. They are the ones making sure that only their families and allies profit, building magnificent estates for themselves. They are the ones who made sure only their people and unions received construction and maintenance work. They are the ones who used the money to tighten their grip on the town.

One man sighs to us, *"Aqui tenemos caciques lamentablemente"* (unfortunately, we have caciques here).

And not only do the caciques keep the money for themselves, they incite violence and the disruption of everyday life by organizing their allies in innumerable strikes and blockades in what seems to all an endless game for influence and money. People speak openly of votes bought and sold (for six hundred pesos an election). Perhaps a few older La Ventosans vote out of conviction, but the great majority of votes, it is said, can be purchased or, more accurately, rented.

The wealth of the dueños also attracts unwanted attention from outside. We hear stories about petty criminals from Juchitán and Matias Romero, even from as far away as Veracruz, coming to town to steal from homes or to break into cars or to plot the kidnapping of dueños. A sense of insecurity abounds, and some say *delincuencia* (delinquency) has been on the rise too.

Victor, a teacher, takes us into his home, and his mother makes us delicious tortillas on her comal. Victor's wife proudly shows us the Zapotecan *traje* (dress) she has been sewing by hand for the past six months. They are not critical of the wind parks per se but of the effect their money has wrought on the community.

"My country is a very nice place, *muy solidario*, but when money is involved, it's like a bomb going off." Victor laughs. "It's like Hiroshima and Nagasaki; everyone comes apart."

FIGURE 2.10. The edges of La Ventosa

Elsewhere, we hear a few truly chilling stories, like the one about an intrafamily dispute over a hectare of land for which a rental contract is being sought. A man is said to have organized the rape of his cousin in order to get her to back away from her land claim.

On the western fringe of town, across the highway, where most of the homes are very humble, some little more than shacks, we talk with families that live closest to the turbines. The wind seems to blow with special ferocity here as, seeking shelter from gust-borne gravel, we run from house to house. Residents tell us about birds killed by turbines that have fallen into their water source and contaminated it. They hope we can confirm or disconfirm local rumors that the turbines might cause cancer and infertility. Sometimes the turbines are described as fans that blow the air more violently, extracting moisture from the soil, rendering it barren. Sometimes the turbines are said to suck the vitality from the air, leaving it weak and unable to scatter seeds properly. In a moment worthy of Cormac McCarthy, a desiccated puppy head skips down the road in front of us, wretchedly driven on by the howling wind. Such are the dark reckonings on the edges of town, fringed with anxiety and uncertainty, turning biopolitical modernity's promises of long life, progress, and abundance inside out.[53]

Winners and Losers

But it must be said, there are still many families that appear quite content with the course and impacts of wind development. And there are others who shrug, saying that there is no turning back now anyway and that La Ventosa will have to make the best of it, come what may.

"This wind," Jose Luis Montaño says, "is hard, and it raises people with strong backs."

As a successful professional tied by blood and marriage to the Monteros, Rusvel is guardedly optimistic. We sit in his courtyard, chatting while our children play. We learn about a lovely Zapotec custom wherein a married couple's names merge, their household being referred to by their given names joined together. Rusvel jokes that we would be called "Mingomena."[54]

Rusvel's veranda is packed full at the moment with dozens of sofas, orphans of a tractor trailer that el norte had blown off the road some miles east of town. Rusvel is one to speak quietly and guardedly about the political conditions in La Ventosa. Like a good journalist, he listens well. And, he mentions occasionally, but only when we meet him in Juchitán, that the reign of the Monteros and the marriage with the Spanish companies is not to everyone's liking. He, too, has questions.

"For example, this paving of the streets. These are fairly simple obras," he says. "They could be aiming for something much more important, like building a primary school. For years, the company has promised to build it, but nothing has happened yet. . . . Another example is the fact that the whole town is encircled by turbines now. And I'm looking at this and saying to myself, How is the town going to grow? There are rules about building homes closer than five hundred meters to the turbines. It's not going to be possible to grow the population much more than it is today."

Imprisoned by their own good fortune, even the supporters feel that something is still missing in La Ventosa. But perhaps the gap between hope and reality is being paid down in installments.

In July 2013, shortly before we leave Mexico, the Bacusa Gui (firefly) cultural center opens in La Ventosa. In the local press coverage there is an image of Daniel Gurrión, mayor of Juchitán, sitting between Porfirio and Hageo Montero in the front row of the ceremonies.[55] Special awards are given to Montero cousins Rusvel Rasgado López and Jose López de la Cruz for their distinguished cultural contributions to the community and for "elevating the name" of La Ventosa. There is no mention of cosponsorship from Iberdrola—whatever contribution they made to the event has been absorbed into the logic of clientelist power. A children's orchestra plays, delicious food is served. This is Guidxhi ra riale bii, where the wind is born and now harvested.

La Ventosa can certainly be viewed as a tremendous success story for wind power globally. A developer in Mexico City once proudly described the area around La Ventosa and La Venta to us as the "densest concentration of terrestrial wind parks anywhere on the planet." From a certain distance, La Ventosa appears to epitomize how a world-class resource can—provided inputs of transnational capital and expertise, with canny government mediation smoothing out cultural and political frictions—rapidly achieve multiple goals: improving local economic opportunities and infrastructure, meeting national energy transition targets and biopolitical aspirations for development, and addressing the global challenges of decarbonizing electricity transmission and remediating climate change. Win-win-win, as we heard so often from wind power advocates in the isthmus.

But our closer examination of the origins and impacts of autoabastecimiento wind development reveals that, even its winningest form, the fantasized perfect marriage of energopolitics and biopolitics is rarely such a simple and agreeable relationship. For one thing, the interests of capital always make it a threesome. And then there is the fact that the greatest beneficiaries of wind power are those who were already winning in La Ventosa. Everyone else—the landless, the jobless—senses the score being run up and their own losses mounting. What actually constitutes the aeolian politics of a place like La Ventosa involves so many agents, forces, and relations beyond the typical stakeholder categories of the PPP contractual imagination like "the government," "investors," and "landowners." As we have seen, aeolian politics in La Ventosa cannot be understood apart from deeper histories of red and green politics in the region and especially the fiery politics of land tenure. The deep mistrust of state and federal governance also permeates autoabastecimiento relations, as does the relative efficacy of caciquismo and the diverse networks of political líderes. With Fernando Mimiaga and his sons, the distinction between government agency and nongovernmental advocacy is blurred by still further forms of patronage and, some would say, corruption. And then there is the electropolitics of the grid, the infrastructural politics of roads and blockades, the class politics and concerns about vital extraction, failing agriculture, and deepening inequality. All this and more belongs to Guidxhi ra riale bii.

We now have a taste of the terroir of the isthmus, finding quite disparate political conditions and fortunes in towns that are only twenty kilometers apart. The next phase of our journey leads us back up the switchbacks of the Carretera Panamerica toward Oaxaca City, where we seek intimate engagement with the more remote constellations of pouvoir—state government, federal government, and transnational capital—that claim to be steering the course of istmeño wind development.

3. Oaxaca de Juaréz

September 6, 2012. For three months we have been searching for the person responsible for managing renewable energy development in the state of Oaxaca. It has involved several conversations, dozens of phone calls and online searches, and visits to nearly every major government building in Oaxaca City. All that has led to us waiting in an empty office in the Oaxacan Ministry of Tourism and Economic Development (STyDE). Not literally empty; there is a desk, a small table, three chairs, and a computer. But there is no evidence that this office has ever been occupied. No nameplate on the door, no books on the shelves, no intimate clutter on the desk. No blinds even to shield us from light and heat of the Oaxacan sun. It all seems a bit surreal, and we laugh nervously as we wait. In the regional news media, the governor and other political elites continue to praise Oaxacan wind development for the hundreds of millions, even billions, of pesos of investment it is bringing into the state economy. And on our previous visits to Oaxaca, Fernando Mimiaga Sosa and his allies left the impression that the Oaxacan state government was a major force in steering the course of wind development.

We wondered, From all that to a vacant office in less than three years? What had happened?

This chapter examines the relationship between the Oaxacan state government (particularly its bureaucratic apparatus and political parties) and wind development in the Isthmus of Tehuantepec. In contrast to the potent and occasionally turbulent powers that we found at work both in the isthmus (chapters 1, 2, and 5) and in Mexico City (chapter 4), the capacities of biopolitical, capital, and energopolitical enablement within Oaxaca City

FIGURE 3.1. The office of the coordinator of renewable energy, State of Oaxaca

seemed limited, even absent. And yet we knew from our experience with Fernando that matters had been otherwise as recently as 2010. The apparent vacancy of state governmental engagement with wind power in 2012–13 presented us with an intriguing and often-frustrating riddle.

What we came to discover is that we had overestimated the influence of state government over istmeño wind development during our preliminary research trips in 2009 and 2010 because we conflated the work of Oaxacan PRI party networks with the authority of certain offices (particularly Fernando's) in the state bureaucracy. Given Oaxaca's history of more than eighty years of unbroken PRI governance, the mistake was understandable. At the time, and in the "institutional-revolutionary" logic of the PRI party-state,[1] party-political influence and the charismatic power afforded to certain well-placed party members was fused as seamlessly as possible into bureaucratic authority.[2] It was an unspoken truth that the party and its charismatic leaders governed the bureaucracy rather than vice versa. That is, until the period of 2006 to 2010, when the scandal-clouded governorship of Ulises Ruiz faced the rise of a strong anti-PRI social movement (APPO) spearheaded by the Oaxacan section of the Mexican teachers' union (SNTE).[3] The violent repression of APPO by the state police force helped to generate the urgency to finally bring a fragmented field of opposition parties and movements together behind the candidate whom Ruiz had defeated in 2004, Gabino Cué Monteagudo. Cué ran again for the governorship of Oaxaca in 2010 against PRI candidate Eviel

Pérez Magaña and narrowly won amid widespread accusations that the PRI had (again) attempted to rig the ballot boxes in their favor. As the apparatus of PRI governance (if not PRI influence) rapidly dissembled in late 2010, there were many political vacuums left in its wake, among them the state government's involvement in wind power.

No One's in Charge

Such is the opinion of Fernando Mimiaga Sosa and his sons. We meet them again in Oaxaca City in the summer of 2012 at a health food restaurant called 100% Natural on the southern end of Llano Plaza. Fernando generally seems well, perhaps a little thinner and more weathered than in the past, but in good spirits overall. We chat about the contemporary state of the wind industry and joke about our previous trips to the isthmus. He and his sons are now working on their fundación (see chapter 2) full time. "It's become more of a business now, of course," he mentions with a significant smile. After he leaves to attend another meeting, the younger Mimiagas, Júlio César and Fernando Jr., make a pitch to us to hire the foundation on as external consultants for our NSF grant.

We explain that our grant is not really of the scale to permit the hiring of outside consultants, but they go over the key points of the pitch anyway. The foundation remains very well connected, they say, both in the federal and state governments as well as with stakeholders in the isthmus. They can help us make those connections, and they have an archive of original documents related to the negotiation of land contracts. Those documents would be a gold mine from our perspective, but this is clearly a pay-to-play offer, which we politely decline.

Fernando Jr. speaks at length of how important the foundation has been to the organization of wind power development. "My father was really the first one to have a vision for what could be done with the wind there. And we put together the first plan on how CFE could expand the electricity grid in a way that would convince investors to come. We presented that plan in 2004 and 2006, and eventually the regulatory commission, CRE, adopted it and made sure that the plan, including the new substation in Ixtepec, happened." The foundation, he says, even became involved with organizing the financing of projects, helping to ensure, for example, that standby letters of credit were available to help attract investors and developers.[4]

Leaning in over the table, Júlio César interrupts his brother's rendition of their legacy to focus on the obstacles wind development faces now. "The problem is that when my father left office, no one took over his portfolio. That is, his portfolio was spread out over, I don't know, three, four, five, six people. And these people don't coordinate with each other, so, basically, no one's in charge. . . . There's no common vision, and that's a problem. Right now there seems to be a major push in the state government on solar energy. I applaud that, of course. They are doing various studies. But we hear it from reliable sources that this government is not very interested in wind energy. Why, we don't know."

It seems that the handoff between administrations was not particularly smooth, at least as far as wind power was concerned. And subsequently the Mimiagas have been almost completely frozen out by the Cué administration (or perhaps they felt it not in their interest to open their networks of influence too generously to the agents of another cluster of political parties).

Fernando Sr. tells us later that he feels he has been made a scapegoat for the incompetence of the current administration, particularly as political unrest around the wind parks intensified in 2013. He says, with a genuine tremor of injury, "Now people say that *I'm* the one who created all these problems down in the isthmus. But you two know that that simply isn't true." What is most striking to us in the summer of 2012, however, is that the Mimiagas seem honestly not to know exactly who took charge of Fernando's portfolio after he left office.

Júlio César waves his hand at the question. "I'm sure you if you look at the [government] website you can find a list of the actors involved. What you really need to know is that they are coming at this with *perspectivas muy partiales* [highly particular perspectives]. They don't have a general vision for the region of the isthmus or for the state of Oaxaca."

Since the Mimiagas can no longer help us to gain access to the state government, we next try a friend of a friend, Gaby Blanco, who is one of Cué's administrative assistants and works with him on a daily basis. We hoped she might be able to schedule us a meeting with Cué himself, but owing to the now-annual summer ritual of teacher's union strikes,[5] the Palacio del Gobierno on the Zócalo is being blockaded, and Cué is working from another location trying to resolve the situation. Instead, Blanco meets with us to share what she has learned in her two years working at Cué's side.

Blanco epitomizes the young, professional cosmopolitanism that we come to recognize as Cué's desired public image for his *sexenio*,[6] in which he has made transparency and accountability—as far as we can tell, quite sincerely—the watchwords of his administration. She is flawlessly bilingual and refreshingly candid about the hopes and impasses Cué faces in trying to undo the legacy of Oaxacan PRIismo.

When we ask her to what extent she thinks wind power development is a specific priority for Cué, she says she can only recall one meeting the governor has had. A delegation from the lagoonal fishing town of Santa Maria del Mar visited Cué to appeal for his help in ending the road blockade of a neighboring community (San Mateo del Mar) over a territorial land dispute. To accomplish this, they asked that Cué publicly endorse the wind farm project whose developer had promised to build a new road to connect them to the mainland. Blanco is somewhat vague on the details, but she is describing what would later come to be known as the Mareña Renovables project.[7]

We ask whether the governor endorsed the project, and she equivocates slightly.

"On paper, they have an ideal model," she says. "But then we also had complaints from San Mateo. This is a place where, let's say, to receive a hundred dollars would be a big deal, and suddenly the company was giving the municipal authority thousands of dollars all at once. The problem was that since then, the authority has not shared with the community how they are spending that money. There's a lot of mistrust and division. . . . The municipal authority and the agrarian authority are pointing fingers. But we later found out the agrarian authority was also receiving money from the company." She shakes her head. "The governor's attitude was, 'Why can't you just work this out? The project looks great. This other community needs its road. Let's get it moving.' I was thinking to myself: If this is just one case, then the whole situation down there has to be a disaster."

And this was the only meeting the governor has had about wind development thus far?

"I'm aware of the list of projects that he has chosen to build his legacy. Wind energy isn't on his list of legacy projects. He does dedicate some time to it. He does talk to investors about the wind sector, and we have agreements with USAID. But I think this topic is moving slowly."

We then ask to what extent environmental projects as a whole are a priority for the administration. Blanco removes from a desk drawer a substantial binder containing the legacy projects and quickly pages through that.

"Originally there were over a hundred potential projects submitted by his ministers. Then he consulted with USAID and UNDP on feasibility and financing and finally narrowed the list down to thirty-two that we're actually working on. The ministers were given a certain amount of time, around six months, to settle into their positions. But by now, all thirty-two projects should have time lines, action plans, and be in development."

The projects are color coded by their states of progress. She traces a finger along the relevant column. Only two projects are colored red, meaning they are stalled in the design phase, "no consultants or engineers hired, nothing." One of them is titled "Desarollo Integral del Istmo" (holistic development of the isthmus).

"I honestly just think this minister has dropped the ball."

She locates three projects on the legacy list that involve sustainability issues, two of them focused on Oaxaca City specifically. The first is to make transportation in Oaxaca City cleaner.

Blanco says, "He'd like to improve air quality, but the real challenge is all the illegal taxi concessions granted by the previous regime. We have *way* too many taxi licenses here. You know they were political instruments, right? Every election year, the PRI would hand them out. . . . Then we also have a project aiming to clean up the two rivers in town and to create green spaces around them, and then there is the big dam project, Paso de la Reina,[8] and that's it in terms of sustainability-related projects."

We say that we imagine it is difficult to balance sustainability demands against other pressing human needs for infrastructure and care. Still, wind development seemed to have a lot of momentum behind it in previous administrations, no?

Blanco agrees, and although she has little to say about the legacy of the Mimiagas, she also seems somewhat surprised at how little attention has recently been paid to wind power. In the end, she reasons that not only had the ministry in charge been lax in putting together a strategy but the official who had nominally been managing wind development was not entirely competent.

Our interview with Blanco ends on a somewhat wistful note—echoing the criticism of the Mimiagas—about the state government's inability thus far to develop a comprehensive wind development plan for the isthmus. "Right now, the wind companies work with the federal government to get their permissions from the various ministries, and they work with the local governments [*municipios*]. But they should be working with us too. That is to say, we should have a plan, we should be helping them."

She thinks there are others in the government who can tell us more about the state's involvement with wind power development, more than the nameless renewable energy director who might or might not already be in the process of being relieved of his duties anyway. Blanco strongly recommends that we speak next with Oaxaca's director of public investment, Adriana Abardia, whom she describes as "one of the most experienced and skilled and competent administrators that we have anywhere in the government."

We meet Abardia about a week later at a stylish, Spanish-run restaurant near Santo Domingo Cathedral, a huge tree dominating its partially covered courtyard. She laughs brightly and often during our two-hour meeting. She was born in Oaxaca but had long given up on Oaxacan politics before Cué came to power. The chance for fundamental reform lured her back after eleven years in Mexico City. Her job did not even exist before her appointment. Now she has been tasked with creating regulations and evaluation mechanisms for public investment in development projects.

"For example," she says, "when I worked for the secretary of water, we were paying for one hundred wastewater treatment plants, but only twenty of them were operational. It was highly political. But that was an unacceptable burden on public finances."

She says she has developed a reputation within the administration for being cutthroat with traditional entitlements. "They call me Kill Bill behind my back," she laughs, "Like I've got a katana for them."

Sipping on a beer, she says that being in government has been the hardest eighteen months of her life. "I *love* this city, and I'm a little romantic and believe some things could change. You know the saying: 'Yo soy feliz, pero no estoy feliz' [I'm a happy person, but not happy right now]. Many days are really bad. But at least the state now has the will if not the tools to improve things."

The problem with wind power, she tells us, is that the state government essentially has no legal jurisdiction over it. "We have no direct investment in wind power at all. And what happens is that companies come to us, or more often, they just go straight to the municipalities, find partners, and settle there. It just leaves us out of the game."

We ask what authority the state has to tax foreign wind companies.

"Essentially none, because we don't have a state law yet to regulate wind power. Currently, the payments all go to the municipalities for land use or to the federal government for permits." In general, she explains, state governments have little authority to levy taxes in Mexico. "Ninety-five percent of our budget comes from a federal disbursement system, which makes us incredibly dependent upon Mexico City. And you only get more [funding] if you collect more taxes for them. So [tax collecting] is what we have to concentrate on."

She supports wind power, she says, "but I also believe the parks are having a serious environmental impact. The oil the turbines use goes right into the soil. Can it be cleaned? Who's going to pay for it?"

She speaks in terms of pareto optimal solutions and game theory. "I do believe in wind power. But what are the available games? Who's the winner, the communities or the companies? Let's measure it. You can think what you want about the people in the isthmus, that they're lazy and just want money from the companies. But they are acting this way because they are in a game."

She likes the idea of community-owned wind parks even more. "We need our people to own things and have their share, not just rent our lands. The government should put capital into it to remove risk. If we had public funds or credits, cheap credits, that would be good. But we can't do a wind park entirely as a public investment; it's not feasible and not right. We need the private sector's sense of efficiency and profitability.[9] We don't have that here." She cites her own pilot program to give her staff performance-based pay as an example. "When they do something well and receive a check, they are very suspicious, 'Why are you paying me more all of a sudden?'"

She likes the idea of Oaxaca entering the industrial supply chain of wind power most of all. "We *have* to get into the supply chain. Otherwise we are just land providers and rent receivers. No more than that."

At nine o'clock, Abardia has to return to work for a couple more hours before going home for the day. We express admiration for her dedication. Before parting, she leaves us with two further impediments that the state faces in developing a better wind power strategy.

First of all, there is a trust issue. "Many people think we are the same corrupt government as before, and to be honest with you, I'm not sure that we're not. I'm not sure we're not corrupt anymore."

And second is the problem of administrative competence. "The coordinator for renewable energy we appointed was a complete disaster. He's gone now, but we have nobody on it."

Logical Creatures

The vacuum surrounding renewable energy in the state government endured a while longer. Another round of phone calls in August finally revealed that a new Coordinador de Energías Renovables (coordinator of renewable energy) had been appointed, Sinaí Casillas Cano. That information led us to the aforementioned empty office.

Casillas had no background in government, and we could find no publicly available information about him beyond his Facebook page, which was dominated by pictures of his wife and children and testaments to his Evangelical faith. It featured status updates like "Love for Christ is not everything, it's the only thing" and "My identity comes from God." But he had also posted, "Blockade a store, a government office, a street, etc.? No! That is a union strategy of struggle. It is blockading the future of our children and the future of Oaxaca. Anyone who can't see that is an enemy of Oaxaca." In the midst of the fiercely contested Mexican presidential election of 2012, Casillas's social media also made no secret of his distaste for the PRI and their candidate, Enrique Peña Nieto, whom he assailed with various posts accusing him of corruption, lies, ignorance, and disinformation.

When Casillas finally arrives, he apologizes for having kept us waiting, squints at the vacant desk, and then motions us over to the small table. Casillas is friendly and speaks frankly, although he relates later that his ten years working as a journalist have left him cagey about making public statements and having those misinterpreted. Like the other members of Cué's government we have met, Casillas seems somehow too young for his position. Nonetheless he demonstrates a good grasp of the difficulties facing him in his job, in particular the need to address in some way the mounting conflicts surrounding wind power in the isthmus. It is a situation, he says, "that cannot be denied."

Shaking his head, he laments the ignorance (*desconocimiento*) surrounding renewable energy in general in Oaxaca and specifically wind energy. "This ignorance affects all segments of society. That is, it's not just an ignorance of less educated people, it's a general state of ignorance. And what has been lacking, obviously, is work by the state government and the companies, timely and strategic work, to raise awareness and create knowledge about what this form of energy could represent for people living in the region." We have heard such sentiments before and will hear them many times again from different voices representing the state and federal governments. Any resistance to wind power is rooted in a lack of understanding or awareness of its potential benefits. Once experts inform the people as to the *beneficios* (benefits), they will inevitably come around.

Yet Casillas is also not an apologist for the status quo. He is not naïve, for example, about inequities in the distribution of benefits in the autoabastecimiento model. On another occasion, he admits to us quietly that the Yansa-Ixtepec model is "probably the best possible scenario" even if he is doubtful that it can ever be realized. For now, he assures us that STyDE has a plan to set things on a better path, to get the state government more actively involved in managing wind development in the isthmus.

As he explains the plan, he picks up a piece of paper and sketches out three squares in a triangular pattern. In one box, he writes the names of actors in wind development (academics, municipios, ejidos, companies, NGOS). In a second box, he writes themes that matter to various stakeholders (social development, legal framework of land tenancy, environmental impact). And in the third box, he writes the names of state and federal government agencies that should be involved in managing wind development (CFE, STyDE, SEGEGO, SAI).[10]

The objective is to create a new state-level *reglamento* (regulatory framework) for renewable energy development whose content would be developed through an *esquema participativo* (participatory scheme). But it quickly becomes clear that the only participants in this scheme will be the government and academic researchers. *Los académicos*, Casillas explains, are the specialists in the themes that matter. "It's not going to revolve around the other groups of actors. That is to say, we don't want debates. We're not interested in more debate because they already debate a lot and in a very disorderly way."

Order is a recurrent theme in Casillas's narrative, as is the *puntual* (fast, efficient) use of time. A Protestant ethic, Oaxacan style.[11] Giving various stakeholders the opportunity to further debate one another, he opines, would only waste time. In order to undertake meaningful action, the government needs facts, presented in an orderly way.

"We have to do very timely content work. The academics will obviously have their opinions about these themes, as do all the other actors. Those opinions will result in adjustments to our laws and to the formation of this reglamento. This isn't going to be some kind of forum promoting renewable energy; this is going to be trabajo muy puntual (very efficient work), whose objective is a carefully structured regulatory policy whose contents will be developed in an orderly participatory way, with a proper structure."

Still, Casillas is convinced that all these academics will be able to channel the various interests and perspectives of different stakeholders in such a way as to allow the government to develop a regulatory framework with broad legitimacy. If the reglamento does not have legitimacy, he says gravely, there is no point at all in going through the exercise.

"We cannot silence those people who are opposed to the project [of wind development]. They are opposing it for some reason, for better or for worse. And likewise, we can't simply listen to those who support the project because they only represent half the story. Those who are in favor have a truth that is very clear and precise to them, but those who are against also have a truth that surely has a *razón de ser*" (reason to be).

He speaks glowingly of the *documentos* (documents) that the esquema will generate, each of which will bring those reasons for being into comprehensible and actionable truth forms; after a while, the plurality of documentation fades, and Casillas shifts to speaking of a single documento, a compendium of truth that begins to sound a bit like another sacred text.[12]

"Right now there are some who are only negative, negative, negative, without a real justification, no? Then there are those who are strictly in favor but never think about the problems. The document we want to generate will have all of this balanced. Honestly, not everyone will like it, this document, this reglamento, but it is the only way that investment will take place in an orderly fashion, that communities will receive benefits and understand what these benefits really consist of."

Casillas even reiterates the Mimiagas' vision of a city of experts rising from the arid plains of the isthmus, a biopolitical vista of white-collar professional opportunities extending the horizons of Oaxacan children. "We can imagine how within twenty or thirty years, the important human resources of these companies, the technicians, will be Oaxaqueños. Even if right now we don't occupy those positions, it's our vision."

The one wrinkle—not an insignificant one—that emerges toward the end of the conversation is that the state government has no funding to undertake its plan of action. Casillas says STyDE is discussing possibilities for funding the plan with USAID, with IADB (the Inter-American Development Bank), and even with the World Bank. "Honestly, it's not just the state administration of Oaxaca that lacks funding. None of the state governments in Mexico would have sufficient resources to take on this type of project."

Still, Casillas ends the interview on an optimistic note. Even though locating funding may be challenging, he thinks that their plan is sound and necessary. Only the regulatory role of the state government can guarantee that community interests and developer interests will be balanced in the long run. Tension will be reduced, benefits will spread to all. Again, his idiom reaches for the divinity of reason and order: "We aren't demigods, but we are logical creatures after all. This is a logical plan and one that we think will work."

Sinaí Casillas Cano exudes conviction and seriousness. We leave his office sympathetic to the spirit of the intervention but also wondering whether the change, if it comes, will be too little too late. With so many wind park projects already contracted and under construction, with the evacuation capacity of Ixtepec Potencia nearly filled, with no funding or actual timeline for the participatory esquema confirmed, could a new state governmental reglamento really combat ignorance let alone retroactively address the tensions that have emerged from existing *conocimiento* (knowledge) of experiences with wind developers, their contracts, and the turbines proliferating across the Oaxacan landscape? Could state governmental regulation truly undo the capitalcentric logic of autoabastecimiento wind development? Could it transform the energopolitical enablements of grid infrastructure and electricity provision in such a way as to allow the government's biopolitical aspirations to flourish?

And more than this, Casillas's hands are tied by the character of Mexican federalism itself, which, as Abardia noted, allows its states precious little financial autonomy. A Mexican political analyst estimated in 2005 that the Mexican federal government collected 95 percent of the total revenue in the country even though state and local governments made 48 percent of public expenditures, leaving "state and municipal governments with a very small margin of action and flexibility for the future definition of their expenditures."[13] Without the right to levy corporate taxes to support their regulatory efforts, for example, Mexican state governments had to either beg Mexico City or, in this case, international aid organizations for the right to proceed.

Casillas, though himself a native of Ixtepec, admits that he has not had the opportunity to personally travel through the isthmus, to the towns affected by wind development, to hear about expectations, experiences, and concerns firsthand. His directorate seems to consist only of himself, which doubtless makes the image of a host of externally funded academics bearing truth to his door so attractive. Even as Casillas speaks of the widespread ignorance of the promise of renewable energy, he struggles against his own ignorance as to what is unfolding in the isthmus, a lack of knowledge evidently owed less to lack of interest than to lack of resources and perhaps to lack of status within the administrative hierarchy.

Having made the rounds of the state bureaucracy, we saw clearly that Cué's government had a relatively weak grasp on the administration of wind power. There were good intentions in Oaxaca City, a commitment in principle to renewable energy generation and distant visions of prosperity associated with wind power. There was also considerable desire to claim jurisdiction over the harvesting of the isthmus winds, to play a salvational role in uniting communities and companies in a common purpose, to exert balance, purpose, and design over a process that seemed particularistic and chaotic from the vantage point of the Oaxacan Valley. It was a tantalizing developmental opportunity, perhaps the best one the isthmus had ever seen. Yet the Oaxacan state, through a combination of legal, financial, and spatial impediments, had scarcely any role in governing it.

Since, however, "the state" has many elements in Mexico as elsewhere,[14] we thought it would make sense to return to the PRI party, whose networks had so successfully brokered the autoabastecimiento development scheme in the first place. To what extent had Oaxacan PRIismo and its PRIistas been able to maintain their hold over wind power even after the rupture of their party-state?

L@s Diputad@s

This question pointed us to the PRI party headquarters in Oaxaca City, an unassuming tangerine-hued compound in the upscale Reforma neighborhood. We are able to meet with the current and former chairs of the Oaxacan state congress's Comisión Permanente de Fomento de Energía Renovable (Permanent Commission to Promote Renewable Energy), Rosia Nidia Villalobos and Francisco García López, both PRI representatives of isthmus constituencies. We know that both Villalobos and García are powerful and controversial political figures who are frequently accused of corruption by opponents, yet they remain resilient in their high-ranking positions within both state-level and regional PRI networks. García acts also as the coordinator of the PRI delegation in the state congress; a little over a year later, Villalobos will become mayor of Salina Cruz, the largest city in the isthmus. Beyond this, the Permanent Commission is the most obvious sign of Oaxacan congressional attention to renewable energy.

Large glossy prints of presidential candidate Enrique Peña Nieto on the campaign trail greet visitors to the party headquarters. We are given bottles of water labeled "61st State Congress" and are asked to wait in a conference room. A short while later, Villalobos, García, and three of their deputies join us. The meeting begins with a slight edge of wariness, but once coffees have been distributed and our intentions clarified to their satisfaction, the conversation becomes friendly and lively.

They explain to us that their commission has existed since 2010 to make sure that the state congress maintains a strong focus on renewable energy development. It has not been easy, however. Villalobos blames the Cué government for a lack of coordination and communication with the congress and federal government and for assigning renewable energy a low priority in the ministry of economic development. This lack of coordination has left developers unsuccessfully trying to inform comuneros and ejidatarios about the benefits of their projects and allowed NGOs focused on "social conflicts" to enrich themselves by "generating inconformity for no reason at all."[15]

It quickly becomes clear that their knowledge of the local political nuances of wind development is considerably greater than that displayed by our other interlocutors in Oaxaca City. One of their deputies, Joaquin, seems particularly well informed. He explains that there are a variety of problems in the isthmus related to wind development that have to be addressed. How sorghum production is declining as agricultural lands are being fallowed in favor of wind turbines is one issue. Discrepancies in rental contracts are another. Joaquin discusses how the first rounds of land contracts set rental rates at less than half of what they are now, which has led to protests and highway blockades.

"So we feel that our initiative has to seek to normalize this. We also have to normalize the kind of assistance that companies give communities. It's been the case that some just give money to the mayor, to the commissariats, maybe ten million pesos in some cases. What they do with the payments is up to them, and some do good things, and some don't. That has generated a lot of inconformity as well."

García interrupts to praise the former governor, Ulises Ruiz,[16] for having done more than any other government in Oaxacan history to help regularize land tenancy, offering to waive fees for land registration to help accelerate the process of defining private-property ownership. The problem, in his view, is that the current government has not taken this area of development very seriously; it has dissolved the powerful position occupied by Fernando Mimiaga Sosa ("not often credited, but a real visionary") and fractured it among several other offices.

García says he is strongly in favor of wind development for how it has brought new wealth to the isthmus. "Look at someone like Porfirio Montero. He bought land left and right, rented it to the companies, and now he is very successful. He even has built his own company, Poder Istmo." But he is willing to admit that there were problems in the early days with land speculation, with foreign companies like Preneal "coming in like coyotes, hoarding land, acting like monopolists, then later selling to other companies." That led to a lack of trust with some communities who did not know with whom they were really dealing.

And, Villalobos weighs in, there have been problems with the distribution of benefits beyond the landowners. "This so-called second conquest that one hears easily and often in the isthmus, it describes encountering Spaniards in the restaurants, in the gas stations, all the effort that has been expended to renovate houses for them, to build new houses and hotels. There is another side, of course. In La Venta, for example, this was essentially a forgotten town. Now it has a main street that has been paved, the schools have benefitted, the new houses and cars, there are people with means now. But not everyone. That is the problem, that it has not been possible to translate those benefits beyond those who own land. The beneficio comunitario [community benefit] is still very small. And it is worth noting that the mayors make a good business off the permits for change of land use and construction."

Truly an istmeño PRIista at heart, García singles out the former COCEI mayor of Juchitán, Héctor Sánchez López, and his network as a group that allows itself to be purchased by companies. "They have an obtuse vision of their role. Instead of using those funds for social development in the community, such 'leaders' just keep the companies' payments for themselves."

Joaquin, echoing Abardia and Casillas, mentions that the question of payments also raises the question of the limits of state authority: "An important issue is that any contract concerning the production of electricity is under federal authority. It's a national question, involving CFE. So we have to be careful not to interfere."

On the other hand, any agreement (*convenio*) regarding social development would be made with the municipios and thus falls under local authority. "[In the state government] we're in a very delicate position, we can't be seen to be invading *la competencia federal* [federal jurisdiction]. So we are looking carefully at federal law, particularly the law on renewable energy, to see whether there are *lineamientos jurídicos* [legal guidelines] under which the state government would have jurisdiction, frankly, the power to promote more wind development in the Isthmus of Tehuantepec. This is still under discussion."

Villalobos informs us that the Permanent Commission has in mind to organize its own set of forums, three of them over the next year. "One to hear from academics, intellectuals, and students; another from comuneros, ejidatarios, and mayors; and then finally a third forum for developers and companies." The first forum will be sponsored by a university; the third by a wind company, possibly Iberdrola; and the second will be organized by García himself and will take place in Juchitán. "These forums represent three distinct visions, and we want them all to participate and to speak with us."

Our interview ended with a warm invitation to attend all three forums. In the months that followed, we bumped into García from time to time in Oaxaca City, and the invitations were repeated. However, none of the forums ever took place. We were curious why until we interviewed another Permanent Commission member some months later, Margarita García García of the Citizen's Movement Party, who let it slip that the Permanent Commission had never even held a single meeting.

For her, the commission "unfortunately did not have a real agenda," nor did it have any funding. She believed that forums would help a great deal, but she doubted the leadership of her fellow diputad@s.

"It's a huge problem that we have so little contact with the citizens there [in the isthmus]. There is so much ignorance, . . . but my sense is that those in the PRI don't really want to know anything, because they are the ones said to be selling the lands and helping the companies to enter without explaining to anyone what benefit the people will get. It's the same everywhere. . . . In the isthmus, more than anything, they lie about how long the contracts are for. These are humble people. And when they see a little money, they want to take it. But it's not until a year or two later when the [turbine] towers go up that they really begin to understand the finality of what has happened."

Highlighting the inefficacy of the Permanent Commission is not to doubt that the PRI party continued to exercise considerable authority over Oaxacan wind development. In the isthmus anyway, the PRI networks were potent, and their caciques and operators were capable of creating and removing roadblocks (sometimes literally) to project development (see chapter 5), especially when they worked in concert with their subsidiary construction and transportation unions. We had also witnessed the efficacy of the Mimiagas' organization at work in 2009 (see chapter 2) and knew that several of their operatives in the isthmus had later been rewarded with important positions in regional government or had gone to work for wind developers. The PRI had been playing the game of corporatist patron-clientage in the isthmus for decades and, even in the face of impressive oppositional networks like COCEI and PRD, they remained skilled at it.[17] Indeed, across the isthmus, Francisco García López, known often by his local nickname "Paco Pisa," was considered an important political promoter and fixer for the wind companies. In the heat of the Mareña Renovables conflict in late 2012 and early 2013, a journalist told us his sources were telling him that García was receiving one to two million pesos *monthly* to help the developer defuse the mounting resistance to their project.[18]

What is doubtful is that, in 2012 anyway, PRIismo found institutions of state government to be their most efficient vehicle for the exercise of political authority over wind power development. Fernando Mimiaga had occupied a position in the Oaxacan governmental bureaucracy, but it is notable that he also exerted significant authority through party and personal alliances in the isthmus. That said, his bureaucratic position had obviously meant a great deal, because he faded into the background relatively soon after being forced from his office in 2010. So perhaps it was the broader rupture in the PRI party-state organization in 2010 that caused his patronage network to fall into disarray. Without the existence of stabilizing transregional party-political channels, the extant legal frameworks, as Joaquin put it, left state government in "a very delicate position," navigating between federal and local jurisdictions concerning electricity, corporate income, and land. The delicacy of state political and financial authority within Mexican federalism only reinforced this condition.

One could interpret the existence of vacant renewable energy offices, curiously ghostly participatory schemes and permanent commissions, and proliferating speculative forum proposals simply as symptoms of transitional or neglectful governmental practice, but they suggest also an important mode of political theater. These seemingly hollow political practices resonate with the destabilization of late liberal political ontologies elsewhere in the world.[19] And in this respect, they say something not only about the structural effects of Mexican federalism and the hallucinatory character of pretensions to sovereignty absent the right of taxation but also about a broader phenomenon that Michael Bobick terms in another context "performative sovereignty."[20] Bobick argues—building upon his analysis of the "unrecognized state," Transnistria—that sovereignty is "not simply an attribute of a state, but a process that is performative in nature. . . . Performance is the defining feature of statehood." One might add that where normative, "legitimate" politics is restricted or limited, the performative dimension of sovereignty becomes all the more intensified. As we have seen, Oaxacan state sovereignty was also phantasmatic in this way, at least in terms of its doings and sayings with respect to wind development.[21]

As wind development accelerated from 2009 to 2012, the weakening integrative capacity of PRIismo revealed just how little sovereign authority the Oaxacan state actually possessed, how little capacity it had to govern what its political functionaries generally agreed was a wonderful opportunity for the second-poorest state in Mexico to make good on biopolitical promises of development in one of its poorest regions. As things stood, if prosperity came, the Oaxacan state would scarcely be credited with it; if alienation and conflict ran rampant, the state would be blamed for its failure to mediate contentious politics and to maintain the rule of law. There was a palpable sense of paralysis that hung over our interviews and conversations in Oaxaca City. From that paralysis came the distinctive proleptic fantasies of sovereign authority discussed above, overcoming "so little contact" with the citizens of the isthmus, generating truthful documents, constituting legitimate rule through the veridicality of expert authority, making sure that la comunidad finally saw benefits equitably distributed. Aeolian politics, in other words, had become a pivotal site for the imagination and performance of Oaxacan state sovereignty in 2012–13.

A particularly fine example of such performative sovereignty in action was the one forum on renewable energy that did take place in Oaxaca City during the period of our field research, the Segundo Foro Internacional de Energías Renovables (Second International Forum on Renewable Energy, or FIER). This event was not only a testament to the state government's efforts to project its "ordering" capacity over the rising tensions in the isthmus— seeking thereby to supplant these with a carefully orchestrated image of inexorable and beneficent technocratic achievement—but also of how the unruly turbulent forces associated with istmeño wind could shred such sovereign performance and lay its fundamental artifice bare.

Of Smoke and FIER

The courtyard behind the sixteenth-century Templo de Santo Domingo de Guzmán is buzzing with activity. Young men in suits dodge around purposefully carrying binders. Young women in silk blouses and slacks chat in small groups nearby. There are perhaps 150 people, many of them well-dressed university students, sitting on folding chairs under a large white tent, casually looking at programs or holding them up to deflect the oblique light of the morning sun. The setting, Oaxaca's Jardín Etnobotánico, is stunning, a magnificent celebration of the botanical diversity and civilizational depth of Oaxaca by artists Francisco Toledo and Luis Zárate and ethnobiologist Alexandro de Ávila. The jardín contains more than seven thousand plant specimens, representing the full climatological spectrum of the state; the garden's design and installations reference the cultural diversity of Oaxaca's indigenous peoples and their pre-Columbian heritage. The imperial heritage of Oaxaca is indexed as well, particularly the stands of *Opuntia* cacti used

FIGURE 3.2. Second International Forum on Renewable Energy (FIER), Oaxaca City

traditionally to cultivate cochineals, beetles that were ground to release their carminic acid, the basis of a prized scarlet dye. In centuries past, Oaxacans were forced to offer that dye as tribute to Aztec emperors and the Spanish crown.[22] Not far from the tent, Toledo's sculpture *La Sangre de Mitla* drips cochineal-dyed red water over a large slab of Montezuma cypress. As part of the iconic templo complex, the jardín materializes—as much as any one space could—the biotic, historical, and cultural texture of Oaxaca as an entity, sovereign and singular.

Everyone is awaiting the imminent arrival of Governor Cué to inaugurate the Second International Forum on Renewable Energy (FIER). When he and his entourage of functionaries are seen rounding the corner of the templo, the organizers quickly begin to assemble into an informal welcome line, which Cué dutifully zigzags across, shaking hands and exchanging greetings. Cué is, as befits his public image, dressed a little more casually than some, in a navy blazer and tan slacks, a light-blue shirt open without a tie. He spies Eduardo Andrade, the director of Iberdrola Mexico, standing next to Porfirio Montero and engages them briskly but warmly. Montero slaps Cué on the back, and after Cué turns away, Andrade slaps Montero on the back. Brotherly affection fills the air. Cué works his way to the front of the audience, several photographers on his heels. Some in the audience hold up their tablets and smartphones to capture an image as he passes by.

There are several short speeches of welcome before Cué's. The governor is introduced by the rector of the Universidad Tecnológica de los Valles

FIGURE 3.3. Gabino Cué speaking at FIER

Centrales (UTVCO), Julián Luna Santiago. The UTVCO is the primary organizer of FIER and, though a partnership with University of Freiburg in Germany, received DAAD support to help fund the event.[23] Luna Santiago proudly announces to the governor that they are welcoming experts in renewable energy from Germany, Brazil, China, Colombia, the Philippines, and Romania. Perhaps seeing double, he informs the governor that the audience is composed of three hundred university students who are studying careers in various aspects of renewable energy. Luna Santiago then speaks of the moral imperative of the Kyoto Protocol, of the looming dangers of climate change, of a planet that has warmed 0.76 degrees Celsius in only a decade. Becoming more animated, he gestures to the assembled international experts in the front row, inventing an entirely new adjectival lexicon for renewable energy.

"We have to listen to our older brothers in adapting to tender energy [*energías blandas*], to healthy energy [*energías saludables*]. Today we work together with youths with new ways of thinking. Above all, we are analyzing this potential that we have around La Ventosa, how we can adapt what we have in the waist [*cintura*] of Mexico. Above all, I want us to stop talking only about how Oaxaca is one of the five states with the greatest solar irradiation and to start saying also that we have this favorable wind, above all, for friendly energy [*energías amigables*], for energy that will not contaminate. We stand here today in the very cradle of Oaxaca de Juaréz to change, to set the foundation, to place a new rung on the ladder of development of this country."

Brisk applause follows, and Cué advances to the podium, radiating his own friendly energy.

After thanking the organizers and assembled dignitaries, Cué launches into a brief prepared speech. Here is the heart of it:

> My dear friends, since the beginning of the industrial revolution, humanity has confronted the degradation of natural resources to the point of risking the equilibria of ecosystems and the future of our next generations. It is thus of the greatest importance that society transitions to the generalized use of clean, renewable energy. In Oaxaca, as others have already mentioned, we are putting great force behind developing the multiple advantages and benefits of wind energy in the Isthmus of Tehuantepec, a geographic zone of world-class strategic importance in terms of wind energy production and development. We are opening new opportunities for investment and employment within a framework of legality and consensus, respecting the restrictions of local ecosystems and the decisions of local communities themselves.
>
> To date, the wind corridor in the Isthmus of Tehuantepec generates more than 938 megawatts across eleven parks, and seven companies have installed 685 turbines covering an area of eight thousand hectares, having invested close to $1,900 million. There are currently four further parks in development that will begin operating in 2013 and contribute a further capacity of 462 megawatts, via 339 turbines, on a surface area of three thousand hectares, with an additional investment of $864 million.

Cué goes on to speak of his government's commitment to respecting all Oaxaca's cultures and identities and traditional ways of life as well its natural resources and biodiversity. In closing, he pauses, looks at the audience for a moment, and then decides to go off script to underscore where his government stands on wind power.

"I know that this is a very sensitive topic in Oaxaca right now. We have discussed it, and the government of this state would do nothing, it would promote no project at the expense of our *naturaleza* [natural environment] if it were to put our communities at risk or our environment or our biodiversity." He pauses again, then continues, "We know that there are certain concerns in some sectors of the population, but we will be the first to be vigilant that this type of investment will contribute to the struggle against climate change and that it will be generous to the welfare of our communities. And obviously we will make certain that it doesn't go against our environment or against our good communal coexistence" (*nuestra buena convivencia comunal*).

More applause follows. Cué has testified to the biopolitical and ecopolitical commitments of the state government; he has offered a portrait of lossless flow between the attraction of foreign investment, the planetary struggle against climate change, the preservation of natural biodiversity, and the improvement of communal welfare. Above all, he has performed the legitimacy and sovereignty of the post-PRI Oaxacan state; the state that listens, reflects, respects; that operates consensually, transparently, and vigilantly. The audience responds. Most seem to want to believe him. We want to believe him.

Cué departs before the panels begin, stopping briefly to hold an impromptu press conference at the side of the tent before waving to the assembly and leaving the jardín quickly. In his absence, something feels lacking, and cracks quickly start to become visible in the image of progress, unity, and legitimacy that the governor so ably conjured.

First of all, an announcement is made that due to the nonappearance of several speakers, the three panels will be condensed into two. The first panel, "Challenges of Renewable Energy in Emergent Economies," bears little resemblance to what has been promised in the program. The representatives from CFE and ENEL Green Power are not there, and the minister of STyDE has sent Sinaí Casillas to read a prepared statement in his place. They are joined by a German professor of management studies, a state diputado for the PAN party (the National Action Party), Isaac Rodríguez Soto representing the Pochutla district of the Costa region of Oaxaca, and Mayor Daniel Gurrión Matías of Juchitán. The surprise guest on the panel is Jonathan Davis Arzac, the executive chairman of the Macquarie Mexican Infrastructure Fund, who is in Oaxaca City to launch an aggressive public relations campaign in support of Macquarie's embattled Mareña Renovables project in San Dionisio del Mar.[24]

The talks begin well, with the German professor saluting Mexico's initiative and assuring the assembly that they are making wise investments in their future. In Germany, with all its renewable energy, he notes proudly, electricity prices are declining.[25]

Casillas then reads from his sheaf of papers, adding more statistics but otherwise staying very close to Cué's inaugural narrative. He laments that only 24 percent of electricity in Mexico currently comes from "air, sun, and water," compared with 76 percent from fossil fuels. "The government of this state considers the development of renewable energy to represent a great opportunity for corporate investment but also a great possibility for many communities in this state who can reach a better form of social and economic development. . . . Renewable energy represents the best possibility for many communities to develop their capacities. . . . The beneficiaries are three: the communities, the companies, and our environment."

Everything about Davis Arzac, from his name to his suit, is *chilango*.[26] But he speaks softly, reverently, almost pleadingly, so that the Oaxacan audience will accept the legitimacy of his project. He identifies the investors, the various permissions the project has received, the many successful negotiations with landowners that have already taken place. He thanks every level of government for "incredible support, indispensable support." He notes the one million tons of carbon dioxide that the project will offset each year. But he reserves his most impassioned commentary for the communities who will benefit through direct and indirect employment, whom his company views as partners, who will also benefit through the sale of energy.

"We are a company that understands that success for a business like ours truly depends on the population feeling comfortable [with us] and that we are respectful of the environment. That is, it would be suicide to try to do something other than this. We wouldn't last even two years in operation! . . . What we need to have is the community by our side, and we can, all of us, live together in peace and in a harmonious manner. We can achieve a common good and make an important leap forward in terms of development in these communities."

So far, so good, but with Mayor Daniel Gurrión's speech, differences of interest become amplified. Gurrión, the scion of a powerful Juchitán construction empire and one of the most powerful PRI clans in the isthmus, is smooth and charming. Although trained as a dentist, there is something about his way of speaking that is reminiscent of a boxer: he embraces, feints, darts, hits, moves on.

He says this is a time when entrepreneurs and people in government need to be talking more about "responsibility." Gurrión gestures to Davis Arzac and says that Macquarie is one of the responsible companies, and he enjoys working with them, even with all the problems they are facing. But what bothers him, he says, is that no one is talking about what municipalities are paying for electricity service. "I am paying, what, a million and a half for public lighting in Juchitán? And everywhere you look around town you see wind turbines. Something is wrong with that." He says that *el tema social* (the social question) has been neglected under the government of Calderón, and he hopes it will be a priority for the next government. "I've been trying to convince the Ministry of Energy to let us have two turbines just for the city to help solve this problem. Can you imagine what it's like having to pay CFE a million and a half? Every month! But talking with the government goes nowhere. So far, they are deaf to us."

Diputado Rodríguez speaks next, and he responds somewhat defensively to Gurrión. In addition to the federal legislation promoting renewable energy development, he says, the state of Oaxaca actually has a regulatory law dating from the final months of the Ruiz administration (in 2010), a law of coordination for the promotion and sustainable use of renewable energy sources. Unfortunately, he says as he throws up his hands, this statute has not been effectively implemented, in part because as a law of coordination, it requires voluntary participation from the companies, the federal government agencies, and the local government, which is difficult to organize, especially when all these partners are thinking solely in terms of their own interests. "But what is certain," he says, "is that we have to create an equilibrium between private enterprise, all three levels of government, [and] society, with individuals, with the commissariats of bienes comunales and bienes ejidales, [and] with local authorities." To this end, he thinks that wind companies' current federal and municipal payments need to be redesigned so that the state receives payment as well "since, in the end, it is us who is responsible for keeping the peace when something goes wrong." He also criticizes the companies for headquartering in Mexico City, which drives all the local tax revenues toward the capital, meaning "all the economic growth is happening there and not in municipalities that need it."

The question-and-answer period becomes a lightning round of further indictments of the government and the companies. Porfirio Montero—who is, aside from Gurrión, the only istmeño involved in FIER in any formal capacity—comments from the audience that the government presence is so little that it "could drown in a teacup" in the isthmus, that despite all the money going to *la federación*, as he puts it, the poor municipios receive next to nothing. Looking meaningfully at Gurrión, Montero mentions that no part of what is paid to the municipios for change of land use ever makes it back to private landowners like him. Calderón is to blame, but so is the mayor of Juchitán.

Gurrión fires back at Montero that it is the law in Mexico that companies have to pay their tax to the municipality in which they are operating, simple as that. But Gurrión then takes advantage of having the microphone to go on the offensive again against the companies, wondering why they do not pay an annual license fee to the city, as a bank or even a poor street vendor would.

FIGURE 3.4. Questions and answers, FIER

Rodríguez speaks up in defense of the government, saying that all these "bad vibes and resentment" about the government are really misplaced. The government is trying, and the Oaxacan government in particular has been innovative with their coordination law, given that they lack real jurisdiction under federal law.

In the next round of questions, people wonder why this FIER event was never advertised in the isthmus and why so few representatives from the isthmus have been given a place on the program. Carlos Beas, the firebrand leader of UCIZONI—a Magonista istmeño activist organization that has been fighting CFE, wind power, and other *megaproyectos* in the isthmus for years—says that "something smells rotten in the isthmus."[27] After ten years of wind development, what benefit are these communities really receiving? Their electricity bills are still very high. "In La Venta there are two hundred turbines, but the school doesn't have reliable electricity. The ejido of La Venta has ceded land to wind turbines that used to provide thousands of tons of food for the region. Now a transnational cement maker gets the benefit of their productivity." His voice rising, he warns of ethnocide against indigenous peoples in the isthmus, particularly the Huaves. He warns of spreading misery. "This is a moment when those of us in Oaxaca, in the isthmus, need to ask ourselves whose interest the generation of wind power is really serving."

A Huave leader of the San Dionisio opposition movement then stands to denounce the Mareña contract as a document signed by their mayor behind the pueblo's back without consultation.

A second Huave man stands, his voice trembling, to ask Davis Arzac exactly how much money he is giving to the federal government, to the state government, to the municipal government of San Dionisio and how much to el pueblo ikojts (the Huave people).

The mood is getting anxious, and the bright young men and women organizing the event are flitting around, shuffling paper, and attending to their smartphones with greater intensity. The side conversations erupting throughout the crowd rise to a level that makes it a challenge to hear the speakers. But Davis Arzac answers at length and in measured tones.

He assures the second speaker, Pedro Orozco, that the community will receive 1.5 percent of the net profits of the sale of electricity and that the company has worked together with "community leaders" to develop a plan of infrastructural enhancements of health, sanitation, and transportation that will benefit the entire community. "This is much more, fifteen or twenty times more, than what could be done for the pueblo without our investment."

Orozco does not seem satisfied. "But to conform with the laws would mean that we would have had to have signed a contract. Is there no hope of stopping this? Do you intend to proceed by force if necessary, bringing violence to our pueblo?"

Davis Arzac says, more forcefully than before, "I cannot agree with the way you are describing the situation. And I can assure you that we have never had, and will never have, the intention to proceed with force."

By now a pall has fallen over FIER; the performance of unity has fully dissolved into accusation, suspicion, and a sense of injury on all sides. Cué's discourse is revealed to be a mirage. Even the supporters of wind are turning on each other, looking for positional advantage, testing strategies for extracting more resources from the others. FIER is perhaps the most authentic representation of the turbulent istmeño wind anywhere in the sunny plains of the Oaxacan Valley. Late in the day, the German professor drily jokes to no one in particular that he is "learning a lot about the politics of energy in Mexico today."

Casillas says little, but shifts frequently in his seat, wincing; the man whose *cargo* (office) it is to oversee wind power development in the state government seems unwilling or incapable of bringing order to its unruly participants.

Over the next few days, the Oaxacan news media that covers the FIER forum publishes accounts that focus on the governor's statement, on the development statistics offered by various experts, and meanwhile wholly ignore the tensions that emerged both among the speakers themselves and between the speakers and the audience. This is not unusual for political coverage in Oaxaca, but against the backdrop of spiraling conflicts concerning wind power in the isthmus, it reminds me of my days studying the East German press, when successful governmental action was the ever-reliable "news from another planet."[28]

Governing the Ungovernable

In front of the civil hospital in Oaxaca City, there is another remarkable sculpture, the *Fuente de las Siete Regiones* (fountain of the seven regions), commissioned by Oaxacan governor Alfonso Pérez Gazga and created collaboratively by architect Octavio Flores Aguillón and celebrated artist Carmen Carrillo de Antúnez, which was inaugurated in 1957. The overt purpose of the fountain followed the logic of the annual Guelaguetza festival, a manifestation of both the deep time of cultural heritage and the unity-in-diversity of the indigenous peoples within a superordinate cultural entity, "Oaxaca."[29] But also like the Guelaguetza, it is hard to ignore the spatio-political message of the *Fuente*. On a central pedestal is a bronze male dancer from the Danza de la Pluma—which is, not incidentally, the closing performance of the Guelaguetza, representing the war of conquest between the Spanish and Aztecs—who towers above six bronze female figures placed equidistantly along the edges of the fountain, each representing in her garments and accessories one of the regions of the state (Cañada, Costa, Istmo, Mixteca, Sierra, and Tuxtepec). The *danzante* (dancer) materializes the political imagination of the Oaxacan Valley, a (masculine) survivor of the deadly conquest, who is still indigenous (Zapotec) to the core and whose cultural power has brought the rest of the indigenous peoples of Oaxaca into an orderly political constellation. The figure resonates with the Oaxacan state we have encountered in this chapter, doing its best to dance its desires into being. Interestingly, though, the feminine figure representing the istmo is lifting her skirt and walking, head held high, out of the fountain, her back turned toward the danzante, as though she is done with his performance. This, too, resonates.

In bringing this chapter to its conclusion, it is important to recognize that however hollow or performative the bureaucratic, gubernatorial, and congressional dimensions of Oaxacan statecraft proved to be, there were other Oaxacan governmental entities involved in the politics surrounding istmeño wind development, although in less visible and often half-willing ways.

For example, in February 2013 we met twice with the Oaxacan minister of the interior, Jesús Martínez Álvarez, whose ministry, SEGEGO, had

FIGURE 3.5. *Fuente de las Siete Regiones,* Oaxaca City

been tasked with trying to resolve the conflicts associated with the Mareña Renovables project, which at that point were reaching peak intensity in several communities around the Laguna Superior. We heard from the minister the by-now-familiar theme of how state government was caught mediating between the jurisdictions of federal and local authorities. It was fairly clear that Martínez found such mediation to be an unfortunate waste of his time. His chief aide pulled us aside to explain that the situation was "very delicate, very complicated," and exhausting for the minister, given that all sides of the conflict "were telling a different story" about what was really going on. In our conversations, Martínez was clearly still working through his own feelings on the Mareña case. He blamed the company for a number of stupid moves, but he felt generally favorable about wind power as a developmental tool in the isthmus, and he expressed his opinion that the environmental and social risks raised against it were being overblown.

Martínez attributed most of the blame for the troubles in the isthmus to two organizations, UCIZONI and APIIDTT (the Asamblea de Pueblos Indígenas del Istmo en Defensa de la Tierra y el Territorio), an organization whose leadership was closely connected with the radical wing of the Mexican teacher's union CNTE 22 (Coordinadora Nacional de Trabajadores de la Educación, section 22). These NGOs, as the minister described them, were led by outsiders to the region and existed more or less to intensify social conflicts. Martínez returned to the NGOs several times in our conversation; they were clearly roving constantly on the edges of his thinking. At once an obvious opponent and a scapegoat for the growing resistance to wind power in the isthmus, the NGOs' militancy and their popular resonance (which was not insignificant) both troubled the mediating work of the state government and served as yet another reminder of the limits of its legitimacy to govern. Both UCIZONI and APIIDTT claimed to support indigenous interests and sovereignty against the mestizo party-political order. As such, they sought to patrol and police the limits of valid political intervention by Mexico City and Oaxaca City into the affairs of the indigenous communities of the isthmus. And given that, in some communities that were opposed to the Mareña conflict, the constitutional recognition of indigenous communities' right to govern themselves under traditional law (according to usos y costumbres) was either being actively exercised (e.g., in San Mateo del Mar) or forcefully proposed (e.g., in Álvaro Obregón),[30] the question of the limits of the sovereign authority of mestizo representational politics and liberal democratic order was unavoidable.[31] The constitutional capacities for indigenous autonomy created a political infrastructure that allowed not only indigenous municipios but also activist organizations like UCIZONI and APIIDTT to challenge the authority of Oaxacan political elites.

Martínez did not seem personally suited to spectacular displays of sovereignty; his performativity was low key and circumspect. He wondered whether convincing the company to help clean up the raw sewage problem in the lagoon would help to appease angered fisherfolk; he wondered whether investing resources in cleaning out the irrigation canals from the 1960s would create enough jobs to siphon away support from the resistance; he mused above all about how to outmaneuver the NGOs to bring seemingly intractably opposed sides together for real negotiations. The NGO leaders had no legitimacy in his mind, although he seemed to respect their tactical acumen. To paraphrase Sara Ahmed, Martínez viewed the NGO leaders as having "too much will," and he frequently accused them of opposing projects simply for the purpose of "being oppositional."[32] He seemed entirely certain that without the blockages created by external interests, he was capable of finding a resolution in which everyone would receive some clear benefit from the project, which would allow it to advance.

Yet, over the next few months, it became clear that SEGEGO's experiments were no more capable than local authorities and party-political networks of convincing the resistance to remove their blockades. As the standoff between the government and the resistance became more intractable and sporadically violent in March and April 2013, it was rumored that the state police chief was advocating for greater use of force to remove blockades and to jail protestors. Martínez, on the other hand, was rumored to be committed to pursuing further negotiations. We knew that the company wanted to avoid bloodshed at all costs, especially now that the conflict was receiving international media attention.[33] This left different ministries within the state government at cross-purposes and seemingly undermining each other.

In the end, the stalemate cost Martínez the second-most-powerful office in the state. He resigned in mid-April 2013 and, in what was likely a parting gesture of frustration, his resignation letter to Governor Cué was published by the news organization *Noticias*.[34] In it, Martínez Álvarez suggests that he no longer has Cué's full confidence, highlighting his experience that "issues already solved had been reactivated by internal interests, many of which sought to injure me but worse still ... [and] are damaging to the government and the population. I do not wish to serve as the excuse why many problems that have solutions are being left unresolved."

That same spring, a Oaxacan political magazine featured a cover image of Gabino Cué boiling in a vat labeled "Oaxaca," which is fired by anarchy and corruption and stirred by an obscene savage, whose loincloth lists, among other affiliations, "CNTE 22" and "parque eólico." In the background, a field of wind turbines can be seen rising out of a distant jungle. The crude caricature offers the other side of the colonial danzante and his mistresses. Here, we find "ungovernable" masculine indigeneity resisting and stultifying the good governmental and developmental intentions of the Oaxacan Valley. The image has many overdeterminations: the centripetal forces of federalism, the infrastructural inadequacies magnifying spatial and cultural distance, the constitutional guarantees of indigenous autonomy, the persistence of communal land regimes, the fierce networks of political loyalty and rivalry, and perhaps even the climatological comforts and serene colonial inertia of Oaxaca City. All these forces contribute to this portrait of the Oaxacan state government as a sacrificial victim to anarchic, corrupt forces beyond

FIGURE 3.6. Gabino Cué in hot water

its control. Somehow these turbines rising from the jungle transmogrify wind power into a wild indigenous weapon, a further mysterious instrument through which the eternal indigenous other can taunt a beleaguered, failing mestizo state.[35]

These were the aeolian politics of the Oaxaca Valley, and their oscillation between sovereign desire and absent efficacy added yet more question marks to the win-win-win calculus being put forward by the agents of biopower, capital, and energopower during the main period of our field research. In the end, despite brief cameos at events like FIER, capital preferred to reside in Mexico City and, to a lesser but still-significant extent, in the isthmus itself. Those who plotted the futures of fuel, energy, and electricity in Mexico also gave Oaxaca City a wide berth. Even those with the resources to bring development, health, and rights to the isthmus found little purpose and traction in the Oaxaca Valley. That left Oaxaca's statecraft very much in the mode of stagecraft, at least as far as wind development was concerned. It was certainly not impossible that with the PRI's return to the governorship of Oaxaca in December 2016, Fernando Mimiaga Sosa, or another like him, would ascend to a position that, in a reformed PRI party-state, would be able to exert more influence over the aeolian politics of the isthmus.[36]

But since that story has yet to unfold, we must now follow the main trunk line of Mexican electrical infrastructure northwest to another valley, Anáhuac, also known as the Valley of México, and to the federal capital, Mexico City. In a national politics largely defined by centralized political authority, we seek the knot of capital, energopolitics, and biopolitics that understands itself to govern the development of Mexican wind power in the isthmus and elsewhere.

4. Distrito Federal

October 17, 2012. The caravan ambles confidently down Tres Picos Boulevard in Polanco, passing none other than Hegel Street, the megaphones on the white, bruised, spray-painted CNTE truck enjoying a rare moment of silence after hours of thundering criticism and demands throughout the heart of Mexico City. There have already been stops in front of the offices of the Inter-American Development Bank, the Mitsubishi Corporation, and Coca-Cola FEMSA. All the leading activists against the istmeño wind parks are in the procession: Carlos Beas from UCIZONI, Rodrigo Peñaloza from APIIDTT, Alejandro López López and Mariano López Gómez from Juchitán, and Isaúl Celaya from San Dionisio del Mar. Our friend Sergio from Yansa is there too (see chapter 1); although obviously not opposed to wind power, he sees a common enemy in the corporate autoabastecimiento projects and thus supports the antieólico resistance quietly, on the side. In addition there are dozens of other istmeños representing the ikojts pueblos of San Dionisio and San Mateo del Mar and from the binnizá communities of Álvaro Obregón, Juchitán, Santa María Xadani, and Unión Hidalgo.

Once we reach the front of the Danish Embassy, an unremarkable building in the slate gray of northern European skies, the San Mateans and APIIDTT hastily unfurl their "No al Proyecto Eólico" banners to either side of the entrance. Beas, in a dark suit and purple shirt, stalks the background, scowling, positioning his photographers and videographer. Others rush to affix smaller handwritten protest signs to the embassy, and loudspeakers roar to life as a series of speakers inveigh against wind development in the isthmus and call upon the Danes to show themselves to receive the caravan's documents

FIGURE 4.1. Protest caravan against Mareña Renovables, Mexico City

of protest against the participation of Vestas, a Danish firm, in the Mareña Renovables project.[1]

Some fifteen minutes later, a willowy representative of the Danish government emerges, and those gathered press in toward him. The tension that has crackled at other stops is less intense in this encounter. Rodrigo, clad in a periwinkle Che Guevara T-shirt, is the bellwether; at times he tenses his jaw and shouts at the representatives who are sent out to listen to their demands. When they do not offer their names, he raises his voice to condemn the lack of respect being shown to the communities of the isthmus. But perhaps placated somewhat by this representative's air of Scandinavian formality and politeness, Rodrigo speaks remarkably quietly, and his hand gestures are economical. The Dane says little but listens quietly and attentively to Isaúl's narration of the threat that the project poses to the community and of the death threats and attacks that have been personally directed toward him by the project's supporters. The Dane thanks them for their statements, receives the documents from Rodrigo with a nod, and disappears back into the embassy.

As the caravan begins to reassemble and to roll on toward lunch, we hover near the journalists who have moved in from the margins to get statements, listening to their dialogues. We know some of them already from other events in Oaxaca City and the isthmus; among them is Rosa Rojas, who has written an excellent series on wind energy for *La Jornada*. But there are new figures as well, even including a few representatives of the international press. A tall man who speaks Spanish with an American accent draws our attention; he is Laurence Iliff, who covers Mexican energy for Dow Jones and the *Wall Street Journal*. We chat with him for a few minutes, and somehow it comes out that he grew up in Santa Cruz, California, and went to the same high school as Cymene. He says that he has been following the Mareña story for a while now and invites us to lunch to tell him more about our research in the isthmus.

Electric Crisis

It is several months before we are back in Mexico City again in May 2013, and by then most of the investors in Mareña have quietly written off their investments in the troubled project. Postmortem analyses from well-placed insiders are circulating though it is still several months before the official cancellation of the project in January 2014. Nevertheless, we take Laurence up on his offer and meet him at Milo's, an upscale bistro in Condesa, one of his favorite haunts in the neighborhood. A long-term foreign correspondent, Laurence has covered virtually every major energy story in Mexico over the past decade and has interviewed key actors throughout the hierarchy of Mexican energy administration, up to and including President Vicente Fox. Though he specializes in coverage of Mexico's parastatal petroleum company, Pemex, Laurence is an interested observer of all aspects of Mexican energy. He finds the conflict over wind development intriguing and says he does not fully grasp where the depth of dissatisfaction is coming from.

"I mean green energy is a good thing, right? Aren't these communities getting anything from these projects?"

He nods as we explain the low land rents; they are no surprise to him.

To the accusations of corruption, payoffs to mayors, and so on, he laughs. "That's how they built Cancún too." And then, more seriously, he adds, "The kickback culture in Mexico is pretty well developed. Ten percent is standard on most big projects. Raúl Salinas, Carlos's brother, used to be known as 'Mr. Ten Percent.'"[2]

While talking through the complex interests and intrigues of isthmus wind development, our conversation keeps circling back toward the carbon center of gravity in Mexican energopolitics. In particular, we find ourselves discussing Mexican president Enrique Peña Nieto's much-anticipated energy reform, which promises controversial changes to the development of

Mexico's subsoil fossil fuel resources. It is a huge political issue in Mexico at the moment and one that is attracting considerable attention in the international press not to mention the hungry eyes of oil and gas exploration and services companies across the world.[3] The prospect of privatizing even part of the petroleum exploration and development on Mexican territory would be a historic event and a terrific windfall for the industry globally. Only fifty-seven wells have been drilled on the Mexican side of the Gulf of Mexico as opposed to the sixteen hundred wells on the US side that yielded twenty-six billion barrels by 2014.[4] Add to that the potential for new enhanced oil-recovery services to boost the flagging production of Mexico's existing fields and the possibility of new pipelines bringing liquid natural gas south from rich shale plays in Texas and Oklahoma. In coverage of the reform, talk of Pemex's inefficiency and technological inability to develop Mexico's petroleum resources is used to justify a new teleology wherein eighty years of petronationalism finally cracks open, revealing mouthwatering profit opportunities to global capital.

The prospect of energy reform brings together this volume's analytics of energopolitics, capital, and biopolitics in a particularly vivid way. Although energy development is often interpreted, including by anthropologists,[5] as a phenomenon driven foremost by capital's relentless search to propagate itself, to call to life its machinic apparatus of productivity, the domain of the biopolitical is also never far removed from discussions of Mexican petroleum.[6] This is not simply because subsoil sovereignty was a critical infrastructure of post-Cárdenas Mexican nationalism; in the neoliberal era proponents of liberalized energy production argued that it would strengthen the national economy through new job-generating foreign investment. Capital, biopower, and energopower are imagined to stream fluidly together to stabilize and modernize Mexico's "carbon democracy." We have heard it before: win-win-win. And where better to examine this confluence than in Mexico City, where the sovereign power and governmental apparatus of the nation-state are most pervasively enabled by the concentration of capital, political institutions, and media?

This chapter examines the aeolian politics of Mexico City, but it will quickly become evident that these politics seem dematerialized, at least in comparison to other chapters in this volume. There is less wind and soil here than elsewhere—protests, yes, but no blockades; expert talk and knowledge abound, but material infrastructures of energic and political enablement are often less visible than one would wish. This is in part a matter of research methods that, mostly of necessity, focused more on formal interviews with busy elites in relatively austere work environments. But it also reflects the magnitude of the epistemic, cultural, and spatial distance of Mexico City from the Isthmus of Tehuantepec (which, in local chilango discourse, was undifferentiated from Oaxaca). Aeolian politics were normally treated as technical, financial, and administrative problems and solutions in DF; it was routinely striking to us how little intimate contact and knowledge our interlocutors had with and of the isthmus, a place whose purported cultural, social, and racial differences were objects of fascination and derision. This is the political terroir of the administrative capital of "Mexican" wind power.

Mexico City is more than just a capital that administrates some preexisting set of people, places, and powers; the city is itself a massive machine for the realization of "Mexico," a boiler room that absorbs all available powers of labor, mind, and materials—resources that are drawn from across the nation-state and from far beyond—seeking to orchestrate and export nation-scale imaginations like "national economy" and "national culture." It is surely true that all capital cities undertake this coordinating and circulatory activity to some extent and that urban elite parochialism greases the engines of universality across the world. But in Mexico City in the year 2013, the artifice of Mexican-ness was acutely visible. As drug cartels and their war machines challenged sovereign lands and borders, creating biopolitical and necropolitical administrations of their own,[7] and as uprisings in southern states like Chiapas, Oaxaca, Guerrero, and Michoacán spawned new autonomous zones beyond the control of police and army, at times it seemed as though it was only in DF, and then truly only in certain neighborhoods like Polanco and Condesa, that the elite neoliberal imagination of Mexico as globalizing nation-state was truly secure.

Laurence helpfully draws our attention away from the much-mediated fragility of Mexican rule of law and state sovereignty and toward the humming engine of the Mexican petrostate, which has for decades now helped to give shape and substance to the nation as a whole. Laurence is not unironic in his diagnostics of Mexican petro-statecraft and clearly enjoys poking fun at Pemex (for their poor safety record and lack of familiarity in advanced exploration and production techniques), at SENER (for "sometimes just making things up" in their energy analyses), and at CFE (for their generally dilapidated state, epitomized by their building on Río Ródano, "where it looks like the air conditioners are going to fall out of the windows on top of you").[8] But, on the whole, he seems to feel that the Mexican petrostate, even absent an energy reform, is more robust than it is often given credit for in the international press.

"Mexico's not in such a bad position," he says. "It's true that production has dropped a lot in the past decade, but it's leveled off now at around 2.6 million barrels per day. That's still more than Venezuela and Brazil." He explains that in Mexico, "even though the Cantarell field has dropped from two million to 400,000 barrels per day, the KMZ field off Campeche is now up to 850,000, and the most important thing is that the oil is super cheap to produce, only three dollars per barrel, meaning that with these high oil prices, they are raking in huge profits. Even with the production drop, 2011 was Pemex's most profitable year on record. And, you know, most people, including me sometimes," he laughs, "like to hammer on Pemex and CFE, but there's electricity in most of the country, and when it goes down, it gets repaired pretty quickly. It's not always super clean and nice, but it works, and it doesn't go out of business like Enron."

More than this, Laurence emphasizes, Pemex brings in huge revenue to support the government. "Pemex has $160 billion in sales a year and is able to turn a $50 or $60 billion profit and gives 90 percent of that back to the government in the form of taxes. Pemex provides 35 or 40 percent of the revenue going into the federal government.[9] And then if [the government] needs cash quickly, they'll force Pemex to float bonds in New York so that they can demand six months of tax payments up front. Basically, Pemex is like a bank for the government. They won't starve them, but they keep them very skinny."

He notes that the Mexican political elite are also trying to think past the petrostate. Extracting such a high percentage of Pemex's profits has hamstrung the parastatal in terms of being able to equip themselves technologically to undertake more complicated and risky forms of oil production such as deep-sea exploration and enhanced oil recovery, which are becoming increasingly routine across the world as older fields become exhausted. Moreover, petro-statecraft is vulnerable to market shifts; production declines are one thing, but if the oil market were really to bottom out, billions of dollars would fail to appear in the treasury, and a state already struggling to meet its responsibilities to its citizens would find itself in dire straits.[10]

Laurence also confirms that climate change is increasingly becoming a theme in Mexican politics, "particularly for Calderón's government; everyone I interviewed there *loved* to talk about renewables." Laurence thinks, however, that the government's exit from fossil dependency will not run through a massive renewables campaign but rather through the gradual displacement of an oil-driven economy by its manufacturing sector. "Manufacturing is doing really well here now," he says. "The car industry alone probably brings twice as many dollars into the economy as Pemex does when you look at all the investment, the jobs, and taxes. It has a really good ripple effect. In Puebla, you know, it's all about Volkswagen. And the electronics industry is doing well too. These industries are relocating from China to Mexico because of the lower transport costs for getting goods to the North. . . . It's just a more sustainable situation than relying so heavily on Pemex."

But this diagnosis brings Laurence to what he views as the real energy crisis looming in Mexico. It concerns not oil so much as electricity. "Mexico really needs to do something about electricity. If you are going to build up the manufacturing sector, you need cheap energy to make it competitive, and right now, electricity rates for businesses are very high in comparison to other countries. The way electricity rates work in Mexico is that for basic electricity, just lights and very basic appliances, there is an extraordinary level of subsidy, 80 or 90 percent." He reiterates, "Basic electricity is *very* cheap in Mexico. But then, to offset that subsidy, higher-end domestic use and commercial use gets very expensive.[11] Basically, CFE relies on overcharging their commercial customers to undercharge their residential customers. But they are losing commercial customers in droves now, which means fewer and fewer people to pay for their subsidies."

The aspiration to provide cheap basic electricity to the entire Mexican population maintained the Keynesian-revolutionary spirit of Cardenismo. A year before signing the well-known decree to nationalize Mexican oil production in 1938, President Cárdenas founded the Comisión Federal de Electricidad with the purpose of "organizing and directing a national system for electric-power generation, transmission, and distribution, based on technical and economic principles, on a non-profit basis, for the purpose of obtaining the greatest possible output at minimum cost, for the benefit of general interest."[12]

Cárdenas did not, however, act to nationalize the private, foreign-owned electricity companies that had been operating a variety of regional power systems in Mexico, with uneven technical success, since the Porfiriato.[13] The first decades of CFE focused instead on expanding grid and power plant infrastructure in order to bring electricity to the many parts of Mexico that the foreign companies had neglected. These companies, particularly Mexican Light and Power (MexLight) and the American and Foreign Power Company (AFPC), were allowed to operate until 1960, when President López Mateos finally decreed the full nationalization of electricity. However, in the 1940s and 1950s, the Mexican government set electricity tariffs low so as to reduce the domestic costs of consumption, a development-oriented policy that led to significant problems for the profitability of the companies.[14]

Mid-twentieth-century PRIismo sought to harness both electricity and capital in service of its populist biopolitics of modernization and development. Yet it created enduring tensions between corporate industrial interests and CFE in terms of priorities in the provisioning of electricity.

Laurence recalls that in the old days, "CFE would just kick companies off the grid with no warning when they were running low on electricity supply." And yet CFE was still regarded as an exemplary utility next to the much-maligned Luz y Fuerza, which had been created from MexLight's operations servicing Mexico City and surrounding areas. It seems that PRIista corporatism and a strong and activist union, the SME, operated largely outside the control of the rest of the governmental apparatus.

Laurence laughs, saying, "Luz y Fuerza was a black hole of corruption. In Mexico City maybe 35 percent of the electricity was being stolen, and the employees were totally complicit in it.[15] I lived in a Luz y Fuerza neighborhood, and the installers would come around and tell you how much you could bribe them so that your meter wouldn't run." Years of heavy operational losses, emergency payments from the federal government to keep the lights on, and constant conflict with the SME resolved to the spectacular conclusion of President Calderón sending out a thousand federal riot police to shut down Luz y Fuerza on October 10, 2009, transferring all operations to CFE, finally realizing the Cardenista program of a unified national electricity grid some seventy years after the fact.[16]

The crucial Salinas-era amendment (1992) of the Ley del Servicio Público de Energía Eléctrica (Public Electricity Service Law, or LSPEE)—which allowed for private sector participation in power generation, including cogeneration and self-supply programs (see chapter 2)—was intended to address, at least in part, industrial and commercial concerns about maintaining reliable, affordable electricity service.

As Laurence puts it, "At some point, the big companies were so frustrated with CFE that they began investing in their own power-generation projects to guarantee service, to reduce costs.[17] But these are mostly big companies involved in building power plants, companies like Cemex. For smaller businesses that don't have that kind of capital, the electricity costs are still pretty crippling for them."

The solution? The movement toward decentralized power production seems unlikely to slow down; indeed, Peña Nieto's energy reform will seek to accelerate it, creating new incentives for independent power production.[18] It also seems unlikely to Laurence that CFE will be allowed to remove its subsidies to residential customers anytime soon. It is a sensitive issue, and

CFE lacks the authority to set its own rates; that belongs to the Ministry of Finance and Public Credit, which is famously, according to Laurence, motivated by "political as well as market considerations." The only real move for CFE is to try to bring costs of electricity generation down and to pass those along to its preferred customers in manufacturing. A carbon-fueled solution beckons, driven by the geography of fuel and the logic of market price.

"I'm pretty sure that LNG [liquid natural gas] is going to be a big part of their plan, Laurence says. "Think of all that cheap gas available now in the US. And Mexico has amazing gas resources too if it could develop them. Mexico uses a lot of LNG already, but it's importing it from *Nigeria*." He laughs. "Can you imagine that? So the Mexicans are working as fast they can to build pipelines to bring all that cheap gas across the border down to Mexico."

An energy correspondent for Bloomberg we later interviewed confirmed Laurence's sense of things, saying, "When I interviewed the energy minister four years ago, all the talk was about renewables. Then, when natural gas prices started to go down, that completely challenged the economics of their plans. Now the only thing people are talking about is gas."

The prediction appears to be coming true. Less than two years later, in March 2015, Laurence reported on the first major foreign investment in Pemex's history. It is a deal with private equity investors BlackRock and First Reserve, who have offered $900 million for a 45 percent stake in a pipeline project to bring US natural gas to Mexico.[19] It appears to be only the first of many such projects to come.[20]

The chairman and co-CEO of First Reserve is quoted speaking approvingly of how the Mexican energy reform has opened up huge opportunities for international investors in energy infrastructure projects: "If you take pipelines, for example, [all of] Mexico has 10 percent of the natural-gas pipelines that the state of Texas has."

In 2013, Laurence told us that the electricity crisis probably did not bode well for wind power. "Per kilowatt-hour, wind is still pretty expensive compared to natural gas or coal so I'm sure there are some people over at CFE who are hoping they don't have to incorporate too much of it."

La Comisión

From our time in the isthmus, we are already aware that CFE's affects concerning wind power are complex and at least somewhat contradictory (see chapter 1). We are eager to learn more about them during our time in DF.

And, fortuitously, the closed doors and evasiveness we faced in Juchitán are replaced by many representatives who seem more than willing to explain the exigencies of the grid and electricity to us.

We begin in the first of many CFE buildings around Mexico City, this one a modern office complex on the Periférico Sur. We meet with Francisco Díaz and Francisco Barba, both of whom work for the Direccion de Proyectos de Inversion Financiada (Directorate for Financed Investment Projects), focusing on the environmental impacts of CFE's new power plant developments. Interestingly, we talk less about climate change and the impact of different types of energy development on various animal and plant species and more about how wind power impacts the environment of the grid. The Franciscos seem hesitant about wind power, largely because of the problem of *intermitencía* (intermittency).

"OK, the principal problem with this form of energy, la eólica, is that from the perspective of the Administration of Energy, there's a type of energy we call 'baseload' [*energía base*]," Francisco Díaz explains. "Above all, this is thermoelectric energy where you supply it with combustible fuel, and it works just like a motor. It's *muy constante* [very constant]. Geothermal is that way too. But with all other sources, even hydroelectrics, output will depend on the rains of the previous year. So in a dry year, you might not have enough supply to meet demand. It is the same way with las eólicas [the turbines]. You might have a peak supply and an average supply. But those are just numbers. In reality, the supply is constantly fluctuating between highs and lows. And for that reason, in most cases of wind energy, you need to build extra thermoelectric support installations just to guarantee the supply of energy, a constant flow of electricity."

"Baseload" is a thermoelectric imaginary, one that has coevolved with the fossil- and nuclear-fueled infrastructure we know as "grid."[21] Baseload thinking naturalizes a situation of endless, constant electrical supply equilibrated to endless constant demand.[22] It gives voice to the energopower of steady thermoelectric generation, feeding a stable, efficient infrastructure of current, all conducted with a capital-centered market imaginary tightly wrapped around it like insulation.[23] This logic does not suffer intermittency lightly; any risk or disturbance of flow is viewed as a threat.[24] Thus, not only in Mexico does one find the perverse argument that there should be more fossil fuel installations for each kilowatt-hour of renewable energy brought onto the grid to guarantee reliability.[25]

We probe deeper, asking about Mexico's ambitious clean electricity targets.

"Yes, well, for the most part," Díaz says, "this was a political decision of the government, *de alto nivel* [at a high level], and the Ministry of Energy [SENER] defined it. Those of us who work for CFE, it is our job to carry out what SENER tells us. But we're not in a position to explain to you the logic behind their decision. To be clear, we're OK with the decision, *nos encanta eso*, [we're delighted by it], but we can't explain the reasoning to you."

They do not sound encantado but rather troubled about the charge that has been given to them.

When we ask about the Kyoto Protocol,[26] they note matter-of-factly that Mexico is among the Annex B countries, which are not required to undertake emissions reduction. They say that only five or six hundred of CFE's ninety-eight thousand employees work on environmental issues. It is not a central consideration of la comisión's work because of the existence of the Environmental Ministry, SEMARNAT.

When we ask about how CFE seeks to reconcile local environmental concerns against the global environmental concerns such as climate change, Fernando Barba mentions the small amount of the Mexican population still not served by electricity (less than 2 percent). "Because, as Hernán Cortés once said, our land is like wrinkled paper, there still remain distant communities where it is very difficult to serve them conventionally. And in some cases, solar panels have gone in to guarantee a certain degree of development, to allow them to power their radios and television, even a motor or a refrigerator."

It sounds as though Barba is about to say that renewable energy has important applications in such outlier cases, but it turns out that his point is exactly the opposite. "And [the people in these communities] they say that they want energy that serves them [constantly], sometimes they protest up to the highest authorities that they want *true* electrification." Even on the fringes of Mexico, in the areas most in need of development, Barba argues, people will reject renewables in favor of the truth of the grid.

Jesús Ortega works in one of CFE's oldest buildings, the very one on Río Ródano with the precarious air conditioning units. A half-century ago this was the primary administrative center of CFE; in 1960 it was radiant and festooned with a bright-red banner, seven stories high, celebrating the nationalization of electricity. Its edifice is now predictably more weathered, its interior faded, epitomizing both the resilience and the exhaustedness of Cardenismo. And yet the building brims with activity.

Ortega is a subdirector in Coordinación y Distribución (Coordination and Distribution), which helps ensure that CFE's clients receive the energy

FIGURE 4.2. CFE administrative center, 1960

they need. He apologizes that his group in Distribution is only tangentially involved in renewables and mostly in the form of solar photovoltaics. Now CFE has plans to extend the grid to all communities of more than fifty people (it used to be one hundred people). For smaller communities, they have tried a pilot plan to set up solar PV (photovoltaic) panels to provide "basic energy." But, shaking his head, he too relates that *la gente no tiene confianza* (people lack confidence) in this form of energy. They were angered when the PV panels ceased to function and blamed CFE for it. And then there is the question of cost. He estimates that a kilowatt-hour generated by solar energy costs five or six times that of a kilowatt-hour produced by conventional carbon sources. "We're not in the position to pass those costs along to the consumer as happens in other countries."

His colleague, Mauricio, drops in, dressed in a dapper tan suit. After listening for a while, he interrupts to say that renewable energy is most certainly "a pillar" of the national energy plan, but it is really part "of the policy of expanding electricity generation and coordinating participation in investments to the expansion of electricity supply." By this, he is referring to the temporadas abiertas [open seasons] that CRE and CFE have created to expand private participation in electricity generation.[27] That has led to 1.5 gigawatts of wind power in the isthmus already, and he believes it will go up to four gigawatts when the program is complete. "It brings together the vision of the state with the vision of the *permisionarios* [permit holders]." He cautions that the problem of intermitencia means that it will be impossible to envision an electrical grid focused "one hundred percent" on clean energy. Still, Mexico has renewable resources that should be developed, and he speaks approvingly of the *consciencia social* (social conscience) demonstrated by countries such as Germany, saying that it is sometimes worth the higher cost for clean energy. "The world has to use less hydrocarbons, and that is true for Mexico as well. We're already a net importer of liquid natural gas."

Both Jesús and Mauricio agree that we need to speak with people in the Directorate of Planning who assess future demands on the grid and plan extensions and improvements accordingly. A few interviews later, we find ourselves in a boardroom at the new CFE headquarters on Paseo de la Reforma speaking to Ramón Villagómez Altamirano. Maps of the grid and CFE generation plants adorn the walls. Villagomez is extraordinarily kind and generous with us, staying long after the appointed end of the meeting to make sure we understand CFE's relationship to renewable electricity generation.

Villagómez is worried, above all, about the cost of Calderón's 2011 law that mandated that Mexico have 35 percent of electricity production coming from nonfossil sources by 2024. "We have to be aligned with these national goals, but it has a huge cost, a *huge* cost," he says with a certain degree of resignation, "and I want to be clear that these goals are political, not technical. There is no way CFE would have recommended this."

About 22 percent of Mexico's electricity capacity comes from hydropower, and that is considerable; however, only 4.6 percent comes from all other renewable resources combined (including nuclear power). A radical expansion of wind power is one scenario, and he says that there are three areas in Mexico with excellent technical wind capacity: Baja, Tamaulipas, and of course the Isthmus of Tehuantepec.

"But even if we developed all of that wind capacity," he says, "even with our best efforts, we couldn't meet these targets. There are natural restrictions." He says that CFE has explored other scenarios, and they feel that the only way to conform to these political goals would be to build several new nuclear power plants. But "in our country, as in most countries in the world, people reject nuclear energy. There is a lot of seismic activity, and we'd have huge problems if we decided to build a nuclear plant near a big city."

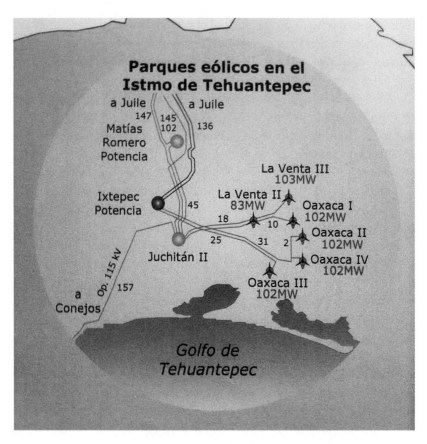

FIGURE 4.3. Map of the istmeño grid, CFE

According to CFE's calculations, meeting the obligations of the 2011 law via wind power alone would cost an extra $3.79 billion annually and $2.4 billion even if nuclear energy were included. Villagómez says that they have thus concluded that "the plan is unrealistic and not economically viable" and that they are lobbying, together with SENER, for a change to this legislation in Congress, aiming at a more gradual transition, with 2050 as the target. They are hopeful that President Peña Nieto will be more receptive to this message than President Calderón was.[28]

But, we ask, does CFE feel threatened by the energy reform, which seems aimed toward further advancing the private-public partnership model and opening the possibility of allowing the private sector to sell electricity as well as generate it?

"Yes, it is a threat. In 1992 the federal government decided not to invest in further expansion of electrical infrastructure by itself. All further expansion would come through working with private partners. And in only twenty years, private generation is now one-third of the national supply. CFE, meanwhile, has no more money for grid expansion." Villagómez seems pained by this admission, which, in all honesty, comes as something of a revelation to us even after years of studying Mexican wind power. With no more public funding dedicated to grid expansion, Mexican electropower had already been made contingent upon flows of international capital in ways that rehearsed the contemporary reform of Pemex and petropower.

Villagómez says thoughtfully, mournfully, "We are worried about the privatization of electricity even if they can't commercialize energy yet. The federal government decided to invest in social development rather than in our growth potential, but how can we achieve growth without federal resources? So, we're very aware of this problem, and one idea is to ask the government for independence, because we are so highly regulated that we can't compete with the private companies. What we say is we want liberty of management."[29]

His statement captures CFE's paradox in a nutshell. The federal government wishes to tether CFE's capacities of energopolitical enablement to the state's biopolitical mission—allowing social development through electrification, guaranteeing highly subsidized electricity to much of its population—at the same time that it is also sacrificing CFE's autonomy, infrastructure, and future to attract foreign capital.

Pemex's future was also being threatened in the context of the energy reform, "but there is a very big difference between these two parastatals, CFE and Pemex," Villagómez says, laughing. "Pemex pays huge, *huge* taxes to the government. So the government treats us very differently."

In Mexico City we learned the extent to which CFE's engineers and administrators felt victimized by the invasion of "political" motives into the "technical" world of the grid. The talk of "market" and "economy" was something of a suture meant to pin together the two otherwise seemingly incommensurable worldviews. But it was clear that CFE's employees felt deprived from full participation in market logics as well. Their concerns about the economic costs of renewable energy transition were not being taken seriously by the government; they were blamed for greed and inefficiency by consumers but not even allowed to set their own tariff rates. Calderón had just authorized a change to the tariff system that had cost CFE a further four billion pesos per year.[30]

La Comisión—so hated and feared for its power throughout the isthmus—looked much more precarious in DF, especially on the eve of energy reform. Everywhere their expertise seemed called into question, their efficacy inhibited. The future of the grid that sustained them had been taken away from CFE. We came to wonder whether CFE's intransigent ambivalence to wind power also had something to do with that forestalled future. Wind power might very well become a significant part of Mexico's electrical future, but none of that energy would be supplied by CFE; it would not be part of their "growth potential." Instead, private producers would profit while CFE would be left managing the engineering challenges of maintaining a grid that had to cope with increasing intermittency. Aeolian politics in Mexico was part of the weathervane pointing toward CFE's diminishment and perhaps eventual abandonment. For once, it seemed we had encountered a win-lose proposition.

SENER

The advice from CFE, that they could not illuminate the political rationale behind Mexico's commitment to an accelerated renewable energy transition, led us to the Ministry of Energy, SENER. It was an especially tense time. The shockwaves of the Mareña fiasco and the Ixtepecan amparo (see chapter 1) were beginning to be felt in the Ministry of Energy during the summer of 2013. Meanwhile, the political situation in the isthmus was becoming very grim. In the space of just a few weeks, new blockades had been set up against Gas Natural Fenosa's 234 megawatt Bií Hioxo project in the Playa Vicente area of Juchitán; the offices of the antieólico community radio station, Radio Totopo, had been attacked; leaders of the APIIDTT and APPJ (Asamblea Popular del Pueblo Juchiteco) were receiving death threats, and some went into hiding; there were even reports of "Zeta-looking" toughs covered in tattoos making the rounds and threatening people who were opposed to the wind parks.

In the end, we were able to arrange two meetings with SENER's new deputy minister of electricity, María de Lourdes Melgar Palacios, and her assistant, Ana María Sánchez Hernández, the first of which was brokered by Sergio Oceransky, who was actively lobbying Melgar to support the Ixtepec community wind project.

Melgar begins the first meeting rather apologetically, saying, "Normally SENER concentrates more on long-term and big-picture energy planning, but when I arrived in office I found these very serious social problems in the isthmus that needed to be addressed." She senses growing opposition to the projects and knows that there needs to be more dialogue with the communities. But her question is how to achieve it. Most of the remainder of the conversation is spent with them asking us to share our impressions and recommendations based on our field research.

Sergio chimes in occasionally, advancing his agenda, pointing out that the previous government allowed these conditions to fester through a lack of oversight and regulation of the whole industry: "To take Playa Vicente as an example, the former mayor of Juchitán, Mariano Santana, allowed a lot of communal land to be privatized in order that this park project could move ahead. There are a lot of people in DF who knew exactly what was going on."

Melgar replies, "Playa Vicente has been much on my mind lately," and then pointedly criticizes the situation she inherited. "The lack of supervision over the whole process is the most scandalous thing to me. Once these projects get permits to interconnect to the CFE grid, and once SEMARNAT issues the environmental permits, there is absolutely no oversight over how these projects are constructed and operate. It's a scandal."

Melgar feels that both the federal and state governments "haven't been doing their homework" about this development process, and she confides that the minister of energy is "really worried about this at the federal level."

Meanwhile, at the Oaxacan state level, there are different powerful political actors vying with one another to steer the process from behind the scenes. "Oaxaca is in such a strange situation in that you have not one but three governors actively involved in the [wind] development process. One of the other governors approached the minister [of energy] recently and said that he could 'take care' of the conflicts if the minister would authorize him to do so.[31] And the minister said to me, 'Just what this needs, *another* governor.'" Melgar laughs, but rather mirthlessly.

Before we depart, Sánchez mentions their intention to take a weeklong trip down to the isthmus to talk to all sides involved in the conflicts. They confess that they are planning to go undercover, perhaps posing as journalists or as researchers ("like you two") in order to get more truthful responses to their questions.

The image of a high-ranking SENER administrator touring the isthmus in disguise—whether out of concern for truth or safety—says much about the informational, political, and spatial disjunctures between Mexico City and its distant provinces. Notwithstanding, we were very curious to hear about what she learned on her trip and to ask what her plans were for addressing the lack of federal governmental oversight over the wind projects.

Our second meeting, in July 2013, is, sadly, disappointing in its content, with an atmosphere that is by turns enervated and tense. We are kept waiting for more than two hours and instructed (twice) not to use our video camera. Although we hope to get at least an audio statement on the record, Melgar and Sánchez seem very uncomfortable at the idea, so we drop it. Melgar absently toys with her smartphone while Sánchez does more of the talking. The visit to the isthmus, it turns out, only lasted thirty hours in the end, and if disguises were used, that goes unmentioned.

We ask whom they talked to and what they learned.

They answer vaguely that they talked to many different groups and have realized that there is a need to define procedures more clearly and to articulate "the correct form of development. We can't simply focus on the business position because that won't be sustainable for long from a social standpoint. Other groups need to feel that their interests are being addressed as well."

They also learned that there are a great many tensions in the isthmus that have nothing directly to do with the wind parks. Development in the isthmus should be not just about wind development; the government needs to have to a balanced regional development plan. The bottom line is that, in terms of managing wind development, "we don't know which esquema [scheme] will be the most useful." Melgar mentions her personal commitment to renewable energy development and notes that wind development is growing in other Mexican states now—in Baja, in Puebla, in Yucatán—as though to say that even setbacks to istmeño wind development will not deter Mexico's energy transition.

Walking us back toward the elevator, Sánchez speaks a little more freely, giving more insight into how at least some at SENER are thinking about the conflicts. "It's too easy to say that it's just the social movements that don't want these projects. It's also structural. That's why government needs to redefine how the development process is designed and managed," she says. "By consultation, you can't just mean talking to the landowners! You have to consult with the whole community. [And] there has to be a process of the diffusion of information first, and then a consultation process afterward so people can raise concerns and objections. That should be the fundamental structure. It has to go through phases."

This would sound very encouraging to those many istmeños we met who were concerned about—but not necessarily ideologically opposed to—the mushrooming of wind parks across the region. But again, as with STyDE in

Oaxaca City, it felt as though these considerate second thoughts about consultation were appearing too late to help defuse the conflicts brewing around projects that were already being built or already in operation let alone to significantly inflect the dominant paradigm of wind development in the future.

In a hopeful sign, in January 2014, Melgar's office organized a private informational meeting for all the renewable energy firms operating in the isthmus, focused on *el derecho a la consulta indígena* (the right of indigenous consultation), with the objective of "guaranteeing the development of projects with a focus on social sustainability, respect for human rights, and strict observance of international norms." A month later, Melgar was transferred, very likely promoted, to become deputy minister of hydrocarbons for SENER. But SENER's commitment to the principle of consulta appears to be ongoing.

In early 2017, on the page of the SENER website devoted to "Sustainable Development" was the following text: "In relation to the participation of communities in the process of making decisions about the execution of [electricity generation] projects, the Subsecretariat of Electricity promotes strategic actions to guarantee that the operation of the electricity sector in Mexico responds to the principles of sustainable development. In 2014, the Subsecretariat developed timely interventions in the matter of human rights and social sustainability."

The text goes on to explain that the Subsecretariat has initiated an "interinstitutional relationship" between SENER and several other federal agencies to help implement human rights and the right of prior consultation of indigenous communities.[32] Details are few as to what this interinstitutional relationship involves substantively. But the mention of a certain planned wind park gives a clue as to the origins and purpose of this initiative: "In accord, the first stage of a process of indigenous consultation has been realized for the construction and operation of a wind farm project in the Isthmus of Tehuantepec, which will have a generating capacity of close to 300 MW."[33] It sounds like Mareña reloaded, but this time beginning with a consulta and pursuing "the correct form of development."

As time would tell, the project in question was indeed a reformation of the Mareña Renovables wind park, now called Eólica del Sur and planned to be built on land belonging to both Juchitán and the neighboring town of El Espinal. In February 2015, SENER organized a consulta in Juchitán to present the project to the indigenous community and invited, among other observers, the former United Nations special rapporteur on the rights of indigenous peoples, James Anaya. Anaya has published his reflections on the event,[34] which offer a detailed and balanced report on the deficiencies of the consulta.

Although Anaya offers warm praise for the "quality and dedication of SENER's team" in the face of a difficult task, he notes that it is not in the spirit of UN Convention 169 to offer a fully formed project to an indigenous community for acceptance or rejection. Rather, the spirit of Convention 169 is for governments and companies to work together with indigenous communities at all stages of a given project's formation. He views the form of the project as beneficial mostly to the company, and he criticizes the proposed distribution of benefits as "inequitable" and the informational methodology of the consulta as "ineffective."

But Anaya's worst criticism is saved for the attitude of the representatives of Eólica del Sur: "During my visit I perceived that the [company's] staff views the indigenous population as inferior, their traditions and cultural practices as backward and their understanding of a right to communal ownership of the land and the wind as having no place in the modern world that they are promising. Operating personnel do not seem to grasp the political context and the risks of investing in a historically oppressed population, which still seems to be waiting to see the benefits of more than fifteen years of wind development, often times which has taken place under unfavorable conditions for the Zapotec population."

CRE

The last stop in our tour of the federal governmental agencies involved (at least nominally) in regulating istmeño wind power development is the Comisión Reguladora de Energía or CRE, which was established in 1995 to essentially act as the permitting and regulatory agency overseeing the growth of Mexico's private energy sector. Over time, CRE was granted additional authority in the electricity sector, especially in the definition of norms of renewable energy and natural gas use. The agency is widely regarded among developers not only as a regulatory agency but also as a broker and go-between that manages and optimizes relationships between investors, developers, and CFE. As one wind developer explained to us with a knowing nod, "CRE is more of a facilitator than a regulator in the traditional sense."

With fewer that two hundred employees, CRE is not a large agency, but one feels a stark contrast in its offices from those of CFE and SENER. The jeans and short-sleeve shirts at the latter are replaced by business suits for all;

colorful ties and boardroom-worthy cosmetics finish off gender-normative ensembles. We meet with Alejandro Peraza, CRE's director of electricity and renewable energy. On the wall behind his desk are a series of vintage photographs of a hydroelectric dam under construction; on his bookshelves are several model wind turbines.

Peraza is calm and affable and a true believer in autoabastecimiento, waving off any idea that this course of development could be injurious to either CFE or the nation. "Our drivers are two: the high cost of public service electricity and the quality of our renewable resources." At CRE, he says, "we don't believe in subsidies, we believe in business [*negocio*]," and that attitude has led to large investments in renewable energy development without financial incentives from the government. They have even done away with intermitencia, he claims, by creating a "virtual energy bank" which tracks in real time the price of energy added to and consumed from the grid. "When generation drops, when intermittency occurs, I simply go to my bank account and withdraw my 'money' there. But, of course, there is no monetary transaction; it's all a matter of energy."

Although this is clearly not the aspect of intermitencia that troubles grid engineers, it is a shrewd bit of financial wizardry that helps to iron out fluctuations in payments. Peraza thinks highly of Mexico's chances to meet Calderón's clean electricity targets, particularly as they begin to pursue more small-scale projects (less than thirty megawatts) that can be developed quickly, efficiently, and with fewer environmental risks. He predicts that renewable energy will lower rather than raise the cost of electricity in Mexico in large part because it will bring in more efficient international partners and reduce reliance on CFE's expensive public service. He does not acknowledge that the high cost of that public service is also paying for high domestic electricity subsidies.

It is Oaxaca that is worrying him. "There is real trouble in Oaxaca now that can't be papered over. We're hoping to have five gigawatts of wind power in production by 2017 or 2018. It's very intense development activity. And obviously we are worried about the difficulties now." The Oaxacans both fascinate and irritate him. As the conversation proceeds, he gets increasingly reflective and speculative about what to do about *los oaxaqueños*.

It is a cultural and historical problem above all, he feels, driven by their affinity for usos y costumbres, "which essentially means there are zero rules, just whatever they say their customs are." He sputters, "Or they have—some of them have—they'll have a piece of paper, handwritten, that says for four centuries they have owned this land. This society, Oaxacan society, is very committed to their internal rules. They might sign a contract with a company one day and then the next day decide, 'No, we don't want that contract after all.' And on the other side, none of the companies have been prepared for these kinds of negotiations; they think that they can negotiate in Oaxaca the same way they do in New York, but *Oaxaca no es lo mismo, Oaxaca no es lo mismo*" (Oaxaca is not the same).

The quality of the wind is so good, though, that prospective investors continue to arrive even as the political terrain becomes more and more risk-laden. Peraza finds this lamentable and ultimately beyond his jurisdiction to solve. "We have no legal powers to solve the problem of social participation. It is not the same trying to do a wind project in Oaxaca as in Tamaulipas. In Tamaulipas they are fighting to have the parks there, opening the door wide to investors, likewise in Baja California, Zacatecas, but when it comes to Oaxaca, everything has become very complicated."

The problems are, he is sure, ultimately driven by money, by the Oaxacans' desire for money. The prospect of all that investment is creating political conflict and competition among them. He swivels in his chair, musing, "I don't know, perhaps we should just promise to give the money to the women. Studies have shown this is a good idea. The women use payments for public works, for pavement, for water, for drainage. But if you give it to the men, they'll spend the money looking for a prettier *muchacha*."

The Banker

From Peraza, we gain important letters of introduction to the communities of financiers and developers engaged in istmeño wind development. The conversations and settings range from a literal smoke-filled room of masculine corporate privilege at the wind industry advocacy group AMDEE to the bright, airy offices and pitch-perfect Euro-cosmopolitanism of EDF Energies Nouvelles. Above all, AMDEE seems committed to informing us of the unstoppable forward momentum of the private wind sector. They represent fifty-six different corporate interests who have been undeterred by "minor setbacks" like Mareña, which is "just one case among dozens of successful examples," they assure us.

Thomas Mueller Gastell, the Mexico country director for EDF-EN, laments the negative publicity Mareña has generated more openly, "since, through the media, it creates unfortunate stereotypes. That all wind developers are bad just like all Americans are spies and all Muslims are terrorists." Still, Mueller feels optimistic about the future of wind development in Mexico. "The re-sources are so good in Oaxaca that even without government subsidies, they will be developed. The strength of European wind needs subsidies, but not the Oaxacan wind."

By far the most fascinating of these contacts is the Banker, who, at his request, shall remain nameless. Of all our interviews across Mexico—short of Porfirio Montero, the supercacique of La Ventosa—the Banker is the per-son who exudes the most confident sense of command over Mexican wind development. He flawlessly voices the logic of capital—of people, land, and permits bought and sold—with an offhand pragmatism. It is little wonder given his investment portfolio. His institution has invested in a number of istmeño wind parks at different stages of development. He personally com-mands more than $200 million of investment capital in renewable energy projects; he has a staff of a dozen in Mexico City and a network of more than a hundred agents managing community relations in the isthmus. His group's annual profit target is $12 million, which will increase to $25 million in 2016. In some projects he earns a modest 1 or 2 percent annual return on his in-vestments over ten years or more. In others, he can manage a 100 percent return in just three years. It all depends on the nature of the project and at what stage his group enters and exits the investment. The Banker speaks frankly, in detail, and with an occasional wicked sense of deadpan humor.

Unlike Peraza, the Banker knows how to manage the Oaxacans: "The mantra of working in Mexico as a foreigner is *patiencia, dinero* [patience, money]. . . . Otherwise you can quickly go crazy." The only real problem he faces in Oaxaca is the evacuation infrastructure to bring his electricity to market. That is why developers are looking for projects elsewhere in Mexico: the existence of a larger grid and more consumers closer to the point of production. Ixtepec Potencia, he confirms, was financed by private investors like him who sought to sell wind-generated electricity to CFE as part of the first temporada abierta (open season). "Oaxaca is a sparsely serviced part of the country, and the existing infrastructure was already saturated, so we had to build new infrastructure for the 1,800 megawatts of new generation they had planned." And there will still be more infrastructure to come. The second temporada abierta in 2011–12 also required that applicants commit to providing (in total) $150 million of new grid extension and enhancement projects.

"CFE are such bastards," the Banker chuckles. "That included reinforc-ing the ring around Mexico City, something that CFE should have paid for themselves."

Developing a wind park in Oaxaca has, according to the Banker, become a rather routinized process. At least he clearly leaves little to chance when his money is at stake. "From greenfield to operation, it takes nine years. In Oaxaca it's all about land. One megawatt of wind occupies ten hectares. So a two-hundred-megawatt wind park is two thousand hectares. That's three or four hundred people I need to convince."

"Now, land tenancy is a complicated thing in Oaxaca," he continues, "and you'll sometimes find four guys associated with one land plot. I mean, technically it's illegal, but it's common. The point is that you've got hundreds of people to convince, and you have to convince them in the most *human* sorts of ways, so it can get quite expensive. Obviously, you need to have locals doing this. They need to know how these people think. I am completely aware that I am the *absolutely* wrong person to go and talk with these guys." Invoking *Star Trek*, he explains, "I don't speak Klingon, and they speak Klingon. That is, it's not just that they speak Zapotec, it is that they have a cultural outlook that is completely alien to me, and for me to try to engage them would only get me into trouble. So instead I rely on an extensive local team. We have a hundred people scattered across our different projects there. People who were raised in the region, who speak Zapotec; the landowners know their fathers. Even then, it doesn't always work."

We ask the Banker how he finds people for his local team.

"Well typically," he responds, "I don't put my money at risk in the earliest stages, so someone else has already bought some people for me to work with, and then the connections are handed over. The point is that the first six years of project development is just about getting contracts for the land and doing the wind measurements. But you can't get the land without conveying a clear sense of when [the posesionarios] will be making the big money. And the Oaxacans, bless them, don't give a damn about percentages," he explains. "They always want the flat rate. So you have to pay them a bit more up front even if you do better on the back end of the deal. It's expensive to get the land contracted and held, and there's a massive inflationary cycle going on now as the best land gets scarcer. Right now, just to give you a sense, to keep a two-hundred-megawatt project idle costs about $1 million a year in payments."

The Banker normally starts to get involved in projects about four to five years in, when analysts begin to calculate what he terms the "economic viability" of a project. He says this process has become immensely easier since the federal government instituted a universal willing cost (known in Mexico as *porteo postal*, or postage stamp) for use of the CFE grid. The porteo postal means that wherever a park is located and wherever the customer plans to use the energy, there is a straightforward payment per megawatt of transmitted electricity, which is inflation indexed and visibly posted on the CRE website.

"Before this," the Banker says, "you had to pay CFE to undertake a special interconnection study that cost tens of thousands of dollars, and there was no way to predict how they would calculate the marginal cost of transmitting energy. It could be anywhere between one dollar per megawatt and thirty dollars per megawatt. There was no transparency; it was a ludicrous situation."

After economic viability is proven, next comes the permitting phase. "The big permit is the environmental permit, but it's largely bought and paid for. I've heard of delayed permits but never rejected ones. Then there is the archaeology permit, the transport permit, the license from the municipality. That's a very expensive one, you've got to pay for the mayor after all, $5,000 per megawatt."

We talk a little more about the contentious issue of municipal license fees in Juchitán right now. He reminds us with a smile that "in Mexico, mayors only get elected once, and only for three years, so their pension fund can't wait."

The Banker sees little mystery in wind park development, and despite all the talk of conflict in the isthmus, he even doubts that it is all that risky if the process is gone about the right way, that is, if money is invested far enough in advance and wisely to ready the terrain for development and if the Klingons are allowed to speak to other Klingons who understand them and whom they trust.

"It's a clear, tedious, and expensive process." He sighs, very much business as usual.

Given his detailed modeling of its contingences, we expect him to say that financial experts like him are the ones truly governing wind development in the isthmus. But when we put that question to him directly, he shakes his head.

For him, CFE is still the key player. "CFE is by default the driver of this process because they are a technically bankrupt organization with one resource—transmission capacity—which they guard jealously. They give you nothing for free. If you have a self-supply project, fine. Then you will be building new trunk lines for them."

He sits up a bit in his chair. The topic of CFE animates him more than the Oaxacans. "There's a reason why they are behaving this way. You have to understand the economic situation of CFE," he explains. "They have a total income of 330 billion pesos annually. Half of that is spent on subsidizing their residential sector." More than 98 percent of their residential customers, he says, receive "a 70 percent discount off the real cost. That's CFE's biggest investment, really their only major investment. They don't have funds for other investments. The average household consumption of electricity in Mexico is 1.4 megawatt-hours per year. National demand is two hundred terawatt-hours, growing 4 percent a year. A household should be paying four thousand pesos a year real cost, but with the subsidy, they pay two hundred pesos every two months. That's a nice deal. But to do that, CFE has to overcharge their commercial customers." To illustrate his point, he says, "In southern Texas, a convenience store will pay half [for its electricity] what a convenience store pays just across the border."

What is *not* driving wind development in Oaxaca, in the Banker's opinion, is any environmental mission. "What is your motive for renewable energy, anyway? Green? Nah. It's nice, but it's not the motive." He explains, "I see two types of companies. With one type, being green is the goal indirectly. A company like Walmart's board is telling them to green their image by making some investments. But for the majority, tariffs are the real motivation. There is remarkable volatility month to month in CFE's tariffs. It is often plus or minus 4 percent; 7 percent is the record in the past year. And you can't hedge Mexican electricity prices. You only hedge by entering into a twenty-year fixed-price contract. So if your company uses a lot of electricity, and you want to know your long-term costs better, it's a good idea; you'll probably save money."

It is late in the day, and the Banker leans back in his chair, his eyelids heavy, his words slightly slurred because of a fatigued tongue. "Let me give CFE credit where credit is due. If theirs was an efficient model, then it wouldn't be so profitable for me to sell electricity." But he recognizes, more so than others with whom we have spoken, why CFE would also seek to delay, resist, and impede private sector energy provision. "If CFE loses too many of those commercial customers they need to overcharge, if they are lost to the private sector, then CFE will really be in the shit."

Forever an Obscure Object of Desire

In our last few weeks in the capital, we endeavor to track down Oaxacan politicians who are working in Mexico City. We feel anxious to complete the work because the well-known Mexican volcano Popocatépetl has recently reawakened and is ominously beginning to send forth steam and ash.

We meet first with Carol Antonio Altamirano, the former mayor of Asunción Ixtaltepec, a town on the road between Ixtepec and Juchitán, who currently serves as a federal diputado for the PRD party. At his suggestion we meet at the Sanborn's in Parc Delta, which possibly serves the worst coffee in the world. We arrive to find Altamirano, who is dressed sharply in a grey suit, pink tie, and striped shirt, staring at a plate of nearly finished eggs. He is irritated at the waitress for not having cleared them away already. A few snippy exchanges later, he settles back into his public persona. He becomes animated, fidgety even, taking demonstrative sips of water, leaning in closely to enact familiarity.

The wind parks are much on his mind. He says emphatically that although the situation is worsening (*agudizado*), the great majority of the people in the isthmus still want the wind parks. "The people want investment, and this type of investment doesn't have a bad environmental effect. *But* they want their traditions of decision-making respected—they don't want investment through payoffs—they want respect for their land, and above all, they want to receive a significant share of the benefits."

He laments, as do so many of the other politicians we have spoken with, the lack of a government agency to coordinate wind development on the ground in the isthmus. "Instead, it is just the companies handing out information, and of course, it is their vested interest to get people to sign their contracts. There's a lack of transparency between the companies and the communities."

The worst thing, however, is the *actitud de soborno* (bribery mentality) that drives the whole sector: so many projects proceed through payments in cash and trucks to mayors.[35] Ultimately, it is counterproductive. "People in the isthmus are very jealous. When the company gives someone a truck, that offends everyone else. So the more money that flows in, the more protests arise. It's a vicious cycle."

The cycle has deep historical roots, he admits. "The isthmus has been poor for so long that rapid influxes of money break the *tela social* [social fabric]. There have always been social divisions in the isthmus, *always*. People are accustomed to divisiveness in the isthmus. But in the old days, the conflicts were without killings. The wind parks were like the match that enflamed the situation."

Altamirano raps his knuckles confidently on the table, saying that he has just drafted a new law for discussion in the federal chamber of deputies to help improve the situation. It has three parts: (1) to impose a 6 percent tax on gross electricity sales, (2) to send those tax revenues directly to the municipios affected by the development, and (3) to guarantee that those revenues will be used directly for social projects. In other words, it will be impermissible for the funds to go toward indirect administrative costs, thus preventing resources from being siphoned away by politicians for other purposes.

"This plan will not only benefit communities directly, but it should also help put an end to the bribery. Indeed, I think it may even save companies money in the long run."

Interestingly, he seems completely unaware of the existence of the Yansa-Ixtepec community wind park proposal even though it would be sited in his legislative district. After a few minutes' discussion, however, he seems all for it, musing, "Perhaps we need a further reform that will incentivize community wind parks so that for every two parks built by Iberdrola, there will be a community park."

In the weeks that follow, we look expectantly for a news report from Mexico City regarding Altamirano's proposed legislation, but we find no coverage at all. It appears to have died behind closed doors—yet another failed dream of aeolian governmentality.

Our last meeting of this set is with Diódoro Carrasco, the former governor of Oaxaca, who is still a political kingmaker for his PAN party. We approach the interview with great anticipation and a little trepidation. The rumor is that (current Oaxacan governor) Gabino Cué is Carrasco's disciple, that the former owes his governorship to the latter's networks. Carrasco's consultancy is also rumored to be working for various wind park projects in the isthmus, keeping things moving smoothly. Some say Carrasco is even on retainer for Mareña, putting pressure on Cué to use violence if necessary to move the project along.

Laurence Iliff has warned us about Carrasco too: "He did some really oppressive things when he was governor. I remember Vicente Fox telling me after he became president that one of the first things he had to do was to pardon all the people Carrasco had thrown in jail in Oaxaca. There were three thousand people in jail just for eating iguana!"

We are scheduled to meet Carrasco at his consultancy office in the Centro Histórico, not far from the Diego Rivera Museum. The office is strangely ragtag. It lacks a sign of any kind; the gray carpet is worn and rumpled. The air is stale, and fluorescent lights buzz and flicker. We are at a conference table awaiting him when his assistant, Paulina, flutters in to tell us that he will meet us in his private office instead, just down the block. Carrasco has been delayed by a meeting with—wait for it—Gabino Cué. We hurry there, excited by the prospect of bumping into Cué as well. Unfortunately, though, we just miss the sitting governor. But obviously *just*: Cué's cigarette smoke still hangs in the air, and a butt still smolders in Carrasco's ashtray.

This office is more what we had expected of Carrasco given his reputation. It is large, richly decorated, and perhaps unintentionally postmodern in its decorative menagerie of temporal dislocations. To our right is a TV playing *We Are the '80s* at a moderately annoying volume. A copy of Bill Clinton's *Mi Vida* is prominently displayed on the neighboring bookshelf among a host of art and antique books. The table we are sitting at is glass and contains an inset shelf of weathered medals. Next to the couch is a vintage L. C. Smith & Bros. typewriter encased in a Lucite cube.

If Carrasco is the predator he is reputed to be, then he is nonetheless a smooth and charming one. He is a good storyteller, combining colorful metaphors and dramatic tales with occasional poetic and paranoid twists. Leaning back on his leather couch, he says that the isthmus has been "an obscure object of desire" forever, at least since the days of Porfirio Díaz. It was a possible site of what became the Panama Canal. North American and European companies came from the around the world, hungry for the *riqueza* (riches) that the isthmus offered. "It's a very special place both in terms of Oaxaca and Mexico."

But the other side of the cultural richness of the isthmus is that it has always fought against Mexican *mestizaje*. "It has this parallel history of conflict, of being permanently at war. They fought the Arabs who came with the railroads, they fought the French, they even fought [Benito] Juárez. Oh, how they *hate* Juárez in the isthmus." Carrasco laughs. "And then came COCEI, the biggest movement of its kind in Mexico, and they fought the PRI."

In the isthmus they have always been "very autonomous and very warlike," he says, and for that reason, it does not surprise him that there is so much crime there as well. He says those who traffic drugs and persons have deep networks in the region. All the development schemes planned for the isthmus since the Mexican Revolution have not amounted to much. Wherever factories were built, they were lost to political conflict, blockades, strikes.

Still, when it comes to wind power, he is an optimist. "It's such a terrific resource. The government won't keep making mistakes about it. The future looks marvelous," he says confidently.

During his sexenio as governor (1992–98), Carrasco does not recall any talk from the federal government about using wind power as a development instrument. "But under Calderón, things changed, it became a priority, and that's a good thing. Unfortunately, there was a lack of regulation, and companies have done too little to benefit the communities." Ultimately, however, Carrasco places blame on Oaxacan liderazgo, on organizations like UCIZONI that sponge up money, prestige, and legitimacy from international activist groups and attack any megaproject they can.

"They say these projects violate the sovereignty of indigenous peoples, and while some of their points are valid, others aren't. And they never offer a positive alternative; it's always a discourse of negation with them." He says that when his administration was planning the Trans-Isthmus Highway twenty years ago, the same groups were saying exactly the same things.

He notes that the projects that have taken place on mostly private land have fared quite well. Even Mareña might have made it if there had been a little more *mantequilla* (butter) at the beginning.

"Carlos [Beas of UCIZONI] says it's going to kill the fish, but I'd rather listen to a marine biologist," he says, laughing. And then he reiterates his own discourse on deficiency, "The companies need to have better social plans. And the government should have been regulating them more strongly."

Finally, he leans in to share a secret theory about the resistance. "Here is my secret theory that I cannot prove. The parks that have really encountered problems are all next to the ocean. Maybe the problem is that someone doesn't want legitimate development along the shore because that would prevent illegal traffic from using it." He leans back, smiling broadly. "I can't prove that, but that's my hypothesis."

As we get up to leave, it feels as though there is still smoke lingering in the room, a trace also, in the Derridean sense,[36] of the absent istmeño other, standing in for the diversity of Mexicos that are not Mexico City: the regions, the indigeneities, the war machines, the distant dreams, and materials without which the lavish national and cosmopolitan life of DF could not exist.

This chapter has explored the forces and impasses, the powers and contingencies, of energy, infrastructure, finance, and state governance in the nexus of all Mexican flows: the Distrito Federal. Through the granular testimony and experience of key actors involved in the management of biopower, capital, and energopower, the monolithic and harmonic qualities of these ensembles of enablement have been dispersed. We see instead their internal contradictions and dense, turbulent entanglements with one another. They enable and disable one another in complexly knotted formations.[37]

The federal government inhibits CFE's ability to extend its own grid as a means of luring more foreign capital to its imperiled project of national development. Yet the only reason the capital will come is because CFE's electricity tariffs are so high, which is, in turn, an effect of the biopolitical mission of subsidized residential electricity. Foreign capital would seem to be the lead puppeteer of the whole drama, and yet it is being stalked and manipulated by a jealous parastatal seeking to prevent its own dissolution in the process of energy reform. Capital is likewise being challenged by "very autonomous and very warlike" critics, even being bled by its "allies" in a region at the edge of the rule of national law. The federal government might also seem to hold superordinate power in that it has created the policies, both Cardenista and neoliberal, that appear to constitute the rules of the game for the nation's electric future. And yet that government knows all too well the fragility of its petro-statecraft in a time of peak oil and climate change and its inability to outmaneuver or silence the Carlos Beases of the world as they undermine and disrupt the already-strained alliances between CFE and foreign capital.

There was no more common sentiment we heard in the course of research than that the government was not in control of what was happening in the isthmus; its biopolitical promises so often seemed to fall on deaf ears there that even its energy ministers were forced to travel incognito in search of answers. Finally, the grid itself—so essential to Mexico's present and to Mexico's future, channeling hundreds of terawatt-hours a year, with demand only growing—faces a dynamic geopolitics of fossil fuel supply on one hand and the intermittency of renewable energy on the other. This is not even to mention the challenges of infrastructural maintenance and extension and the estimated 15 to 20 percent of generated electricity sacrificed in technical and nontechnical losses, three to four times the equivalent loss in the United States or Europe. The grid was certainly not capable of delivering on the promise of the Mexican postpetrostate by itself.

What future configuration of energopolitics—taking this term now in its broadest sense of "energy politics," aeolian and beyond—could guarantee the national health and wealth desired by government; the expanding profitscape desired by capital; the responsive, lossless infrastructure desired by the grid? That future deserves greater discussion. And to the extent that it involves wind power, we must return to where Mexico's wind is being harvested, where contracts are being signed and contested, where turbines and blockades are rising, where blades and machetes and protest marches are already in motion. We must return to the front lines of tension and struggle over the future of istmeño wind to better understand the aeolian political potentialities that are being set into motion there. Our final journey is thus to Juchitán, the semimythical, autonomous, and unruly political hub of the isthmus, where all the sovereign powers and visions of Mexican national imagination often seem like the drifting smoke and evening incandescence of Popocatépetl, a remote and curious spectacle.

5. Guidxiguie' (Juchitán de Zaragoza)

It would take great deal of time to describe to you the state of immorality and disorder in which the residents of Juchitán have lived since very ancient times. You know well their great excesses. You are not unaware of their depredations under the colonial regime and their attacks against the agents of the Spanish government. You know that during the centralized government they mocked the armed forces that the central power sent to repress their crimes, defeating and causing damage to it, making fun of its leaders, and scorning the local authorities. You have been witnesses to these scenes of blood and horror. —BENITO JUÁREZ, Governor of Oaxaca, addressing the State Congress (1850)

Juchitan contains about 10,000 inhabitants, being the most populous community in southern Mexico. Its inhabitants have the reputation of being a very unruly set, turbulent politicians and revolutionists. In the south, no political movement is made without weighing its opinion in the balance of success, which nearly always turns in favour of the side the Juchitecos are on. They have been in Oaxaca often, as well as in Tehuantepec, enforcing their opinions at the point of the bayonet. —GUSTAV FERDINAND VON TEMPSKY, "Mitla" (1858)

Juchitán, the second town on the southern division of the isthmus, is situated on the left bank of the Dog River, 4 miles from Espinal, and 18 miles from Tehuantepec. It possesses three stores but very few substantial buildings. Its inhabitants, in number nearly 6000, are industrious and laborious; at the same time they are a warlike people, and meddle in all political struggles; which occur constantly in this unhappy country. —M. G. HERMESDORFF, "On the Isthmus of Tehuantepec" (1862)

The *juchitecos* are renowned as the most ferocious, untamable fighters in Mexico when it comes to the defense of their own rights against petty tyrants. They are proud of their unbroken record of loyalty to the causes of democracy, equality and justice throughout the turbulent history of Mexico. . . . *Juchitecos* never call themselves *mexicanos* or even *oaxaqueños*; they are first and always *juchitecos*. —MIGUEL COVARRUBIAS, *Mexico South* (1946)

Tecos Valientes

March 10, 2013. In the bright light and rising heat of late morning, friends and supporters of the Asamblea Popular del Pueblo Juchiteco (APPJ) are gathering in the square that commemorates the Battle of Juchitán on September 5, 1866.[1] That battle remains legendary in the city today: the men and women of Juchitán—many armed with little more than machetes, palos, and stones—rallied to drive an invading force of 2,500 well-armed French troops out of the city, slaughtering hundreds as they retreated through rains and swamps toward their garrison in Tehuantepec.[2] The battle ended the French intervention in the isthmus and also helped to extend an already well-seasoned Juchiteco reputation for fierceness in battle against "petty tyrants" and foreign invaders.

People are gathered in small groups chatting; the mood is upbeat and charged. The APPJ is the newest of the asambleas (popular assemblies) that have formed in response to el proyecto eólico ("the wind project," as it sometimes called), in this case specifically to resist the Gas Natural Fenosa wind park project being built in Playa Vicente.[3] Two weeks ago, on February 25, they inaugurated their resistance by means of a blockade along the highway between Juchitán and Playa Vicente, cutting off one of the major access roads for equipment and construction personnel to move. Today is the asamblea's first major march and rally, and many of the most active members of the resistance to wind development, the *inconformes* or antieólicos, as they call themselves, are circulating. Alejandro López and Mariano López Gómez, who have often been identified in the local media as the leaders of the movement (though they fervently disown the title "leader"), are there talking to a group of men from Álvaro Obregón. Carlos Sánchez, a founder of community radio station Radio Totopo,[4] who is known locally as *Beedxe'* (jaguar), is speaking with some compañeros from UCIZONI, including Nacho, one of Carlos Beas's lieutenants, whom we have met on several other occasions. The organizers of the local #YoSoy132 chapter are present as well.[5] But the majority of those gathering, perhaps numbering two hundred and growing, seem to be normal working-class Juchitec@s—men, women, and children of all ages, many obviously campesin@s, fisherfolk, and obreros.

We talk with the Álvaro Obregón delegation for a while. Although this march is nominally focused against Gas Natural Fenosa, the standoff with Mareña Renovables is still very much on people's minds in Juchitán. A fisherman named José tells us that everything is very calm at the Álvaro blockade now. The company and the police are staying away.

He shrugs, "Maybe they've given up. I don't know."

Nacho shakes his head. He thinks this is all just part of Mareña's strategy. "They won't try force again for a while because they failed last time. They'll try diplomacy instead and involve the politicians. They are already talking about doing a public consulta [consultation], but it's just more deception. If they do it, they'll be sure to conclude there is no opposition to the park projects."

He goes on to tell us that the Mareña conflict has been accentuating a rift within the Cué regime between the allies of former governors Carrasco and Murat. The Murat faction wants a diplomatic solution; the Carrasco faction is reputedly ready to use violence to displace those protesting the park and blockading the access roads. "So we'll see."

Nacho shrugs. He says, "UCIZONI is just working to keep the communities informed and morale high and to alert people not to participate in any false consultas."

Around eleven o'clock the group begins to assemble into parade form. A line of children lead the way, holding up handwritten yellow-and-white signs with slogans such as "We were never consulted and do not want the proyecto eólico. Get out Union Fenosa!"; "We are building autonomía not wind parks"; "Out with the political parties and the leaders who sell our patria"; and "Viva Juchitán, death to the bad government, no to the proyecto eólico." They are followed by a campesino holding a large Mexican flag. Other, larger antieólico banners follow, as do more children with signs. A small red Ford with an enormous speaker strapped to its roof leads chants in Zapotec and Spanish, for instance the ever-popular "Zapata vive, vive! La lucha sigue, sigue!"[6] Some men ride bicycles and motorbikes; some women carry sticks and machetes, others carry children. We are even joined for a while by another flag carrier on horseback.

And so we march through the heart of Juchitán to the Zócalo and the Palacio Municipal, the UCIZONI members handing out colorful posters to passersby, which read, *"No al Saqueo Trasnacional!"* and depict a caricatured red-haired Spanish conquistador, with a Mareña flag fluttering behind him, offering a hand mirror to three istmeño campesinos in exchange for their land.[7]

"And where have I heard that before?" one of the campesinos muses.

FIGURE 5.1. Antieólico march, Juchitán

FIGURE 5.2. Antieólico march, Juchitán

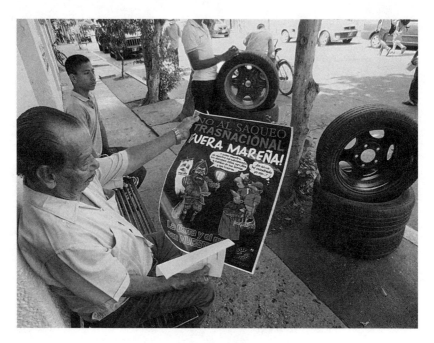

FIGURE 5.3. Anti-Mareña caricature

By the time the parade pauses at the Palacio, its numbers have perhaps doubled. Several speakers take turns denouncing Gas Fenosa and the wind parks in general. But just as much venom is directed toward the political parties and their local leaders, "these cowards and traitors, who have deceived the people and sold the land of Juchitán to the foreign invaders."

The march proceeds into the heat of the afternoon through the Séptima Sección (seventh district), the heart of indigenous, working-class Juchitán, where the men fish and the women clean and sell fish. It, too, is a social space of legend, the crucible of countless uprisings, relentless in its radiation of indigenous identity and sovereignty, a place where, we have been told, "middle-class Juchitecos are afraid to go."

On the way, we discover there is another anthropologist among us, Melanie McComsey, a PhD candidate from the University of California San Diego, who is doing her dissertation research on bilingualism in Juchitán. Melanie confesses that she did not really know what the march was about, but two of her neighbors in the Séptima asked her to attend. She senses that the neighborhood is very divided over the parks. Many of the fisherfolk are strongly critical of the parks, but there are vocal supporters as well, including a woman

FIGURE 5.4. Blockade of the Gas Fenosa project

whose aunt lives in La Ventosa and rented her land to the wind companies and who has instilled a lot of proeólico thinking in her niece. And then there are others who took money from Gas Fenosa early on who now feel guilty about it.

Around two o'clock we arrive at the point on the highway where the road branches off east toward Playa Vicente. About fifty meters before that, we pass five pieces of wood neatly spaced across the road. While the chunks of wood are sizable, they are not so big as to prevent car traffic. This is a symbolic boundary as much as a physical barrier.

Ten meters past that, there are a number of vehicles, some of which are more permanent elements of the blockade, which can be moved into position quickly. There are also vehicular trophies from previous skirmishes with the wind companies in Álvaro and Juchitán, including a large Caterpillar tractor and a white Toyota truck emblazoned with the logo of a Veracruz-based geotechnical services company. Other vehicles are there for the occasion, including one that is dispensing plates of food to the tired marchers.

As the wind whips around us, men take turns hammering away at the truck with palos while the crowd watches and eats lunch. First the windows are broken, then the front windshield is caved in. The front bumper is pulled partly off. There are some calls from the crowd of "*¡Fuego! ¡Fuego!*" but we come to understand from chatting with other marchers that a decision has already been made that the truck is too valuable to burn. At the least, some of the men want to strip it of its engine before burning it. Some boys throw rocks at the truck's

FIGURE 5.5. Battered truck

already-caved-in windows, which ricochet dangerously, earning the disapproval of an older woman, who chases them off, much to the mirth of their friends.

Just when the older men seem to have tired of battering the truck, a wiry young man wearing sunglasses, a tight blue shirt, and jeans rolled to midcalf seizes the largest palo available, a rusted metal bar more than half his height. Wielding it with both hands and cursing in Zapotec, he deals the truck such blows as it has never felt before, launching himself off the ground with some of his strikes. It is a magnificent performance of the *guerrero* (warrior) Juchiteco.[8] A dozen or so swings later, one of the older men decides the truck has had enough and disarms his young comrade, tossing the palo into the truck's bed without further ceremony.

"These Scenes of Blood and Horror"

Both essentialist and constructivist readings of Juchiteco ferocity are tempting. On the one hand, it is not just in Oaxaca City and Mexico City that one hears frequent talk of the fierce, ungovernable istmeños of whom the

Juchitecan warrior is the exemplary figure. This subjectivity is actively indexed and performed locally as well. The image of Juchitán as a city of heroic freedom fighters—home of the last pure Zapotecs, of the Amazon-like matriarchs who invert the patriarcho-colonial hierarchy, of people who have never accepted the imposition of any foreign power whether imperial or governmental or capitalist—is an identitarian discourse that echoes in the restaurants and cantinas, in the mototaxis, at the velas, in the market, everywhere across the bustling city of seventy thousand people. Sometimes it is invoked ironically or with detached amusement; but it is also very often articulated sincerely as a way of making sense of the peculiar political (dis)order of the city, which remains openly, if unevenly, defiant of Mexican and Oaxacan authorities. The Juchitecan political elite today, regardless of faction, seem entirely ready to strike bargains with external powers over land and resources in order to bring themselves revenue and leverage. On the other hand, their status exists in a delicate ecology of neighborhood-level and labor union political networks that militate against national and transnational elite alliances, occupying land and blockading roads as a constant friction against flows of capital and influence.[9]

At the same time, many scholars in Mexico and elsewhere have rightly observed that ethnotypes of lawless and violent indigenous disposition elaborate a colonial politics that validates both the surveillance and repression of indigenous peoples and the law, justice, and modernity of settler governance.[10] As Povinelli has argued, even more recent postcolonial politics of "multicultural" governance—which promise recognition of indigenous subjectivity within the confines of national law and culture—continue to police and hierarchize tolerated and nontolerated modes of indigeneity.[11] Mexico's mestizaje nationalism has exemplified this paradox since the revolutionary Constitution of 1917,[12] a document that promises extensive political entitlements to Mexico's indigenous peoples, including rights to ancestral land and to self-governance. Yet these entitlements have always borne the mark of settler paternalism,[13] have been at best partially institutionalized, and have always coexisted with both violent and nonviolent suppressions of indigenous culture, identity, and authority. Indeed, in a certain sense, Mexican mestizaje nationalism foreshadowed the contradictions of multicultural recognition well in advance of other settler liberalisms.[14] Juchitán's particular position on the fringe of colonial and postcolonial empires helps to illuminate those contradictions and to make sense of how indigenous *resistencia* (resistance) and refusal has come to be made the ethnological soul of the city.[15]

That soul has been many centuries in the shaping according to historical and archaeological accounts. Zapotec presence in the isthmus is generally dated to the mid-fourteenth century as part of a migration/invasion that was likely driven by violent encroachment from Mixtec (and later, Aztec) populations into the Valley of Oaxaca. The Zapotecs, in turn, warred with and partly displaced the Huaves, Chontales, and Zoques, who already inhabited the isthmus. Between roughly 1350 and 1450, Zapotec civilization had two major political centers among its city-states, with the fortress at Guiengola (just north of contemporary Tehuantepec) coming to rival and eventually surpass Zaachila's prominence in the Oaxacan Valley.[16] In 1490 the Zapotec lord Cosijoeza moved his capital from Zaachila to Guiengola, in part at least because of pressure from Aztecs and highland Mixtecs, which completed the transfer of political and cultural authority to the south.[17] Still, the pull of the isthmus was likely as strong as the push from the valley. The resource-rich southern isthmus was already a trading center, and Tehuantepec had established economic importance, supplying the rest of Oaxaca with salt. Historians and archaeologists of precolonial Oaxaca suggest that the Aztecs' move to dominate the Oaxacan Valley at the turn of the sixteenth century and to extend their empire as far as Tehuantepec was motivated principally by a desire to guarantee control over the Central American trade routes passing through the isthmus.[18] Still, it is remembered in Juchitán that Guiengola never fell to the Aztecs' siege, and only a marriage between the Aztec and Zapotec ruling lineages ended the conflict. Only two decades later, the Spanish invasion rapidly unmade the Aztec empire, and so it remains the local sentiment in the isthmus that they were never truly subjects of the Aztecs.

The Spanish arrived in the isthmus soon thereafter, chasing gold and war. Hernán Cortés's brutal lieutenant Pedro de Alvarado arrived in Tehuantepec in 1522—fresh from leading the destruction of Tenochtitlan less than a year beforehand—and organized the Christian conversion of the last Zapotec kings of Guiengola—Cosijoeza and his son, Cosijopii—along with nominal Zapotec acceptance of Spanish hegemony. But the local authority of Zapotec caciques remained relatively undisturbed insofar as they were willing to assist the Spanish in their efforts of wealth extraction; Cosijopii, for example, remained the cacique of Tehuantepec until his death in 1563.[19] Cortés quickly identified the isthmus as one of the most potentially profitable parts of New Spain, and he maneuvered to have the encomiendas awarded to him when he was granted the position of Marquis del Valle centered there. In addition to gold and silver mining, by 1528 Cortés had built in Tehuantepec the largest shipbuilding enterprise anywhere in New Spain to help export the livestock raised on his several *haciendas marquesanas*.[20]

As elsewhere in the Spanish colonies, disease, corvée labor, resource extraction, and Christianity had massive effects on precolonial cultures, forcefully dissolving precolonial social institutions; challenging worldviews, cultural practices, and languages; and helping to generate *el indio* as a composite subject of salvation, labor, and violence.[21] Based on records of indigenous tributaries to the Spanish crown, John Tutino calculates that there was an 85 percent decrease in the indigenous population in the isthmus between 1520 and 1570, a population that would continue to shrink until the middle of the seventeenth century.[22] Yet, at the same time, the Spanish cultural presence remained limited in the isthmus, which reduced the pressure of assimilation. As many of the Spaniards' colonial economic ventures failed over the course of the sixteenth century, the direct exploitation of istmeño resources abated as well. Tutino writes:

> By 1560 the force of the conquistadorial incursion into the Isthmus had diminished. The placer deposits of gold were not generating profit. The shipyards were closed. Other interests guaranteed that Acapulco would become the principal Pacific port north of Panama. For a time, heavy cargos were transported ocean-to-ocean across the Isthmus, but in fifty years, even this previously important sector became a vestige of its former self. The region became increasingly marginal, and the heirs of Cortés renounced their legal rights over Tehuantepec almost without protest.
>
> Following the abatement of the conquistadors' intervention in the Isthmus, a limited Spanish presence and a much-diminished indigenous population remained. In the city of Tehuantepec there was a royal judge and some Dominican friars, and they were surrounded by a small Spanish community. In 1598 this enclave included only twenty-five families. It grew to about one hundred families in the first decades of the seventeenth century but never supported a very large Spanish population. In 1742 there were no more than fifty Spanish and mestizo families in all of Tehuantepec.[23]

Over the course of the seventeenth and eighteenth centuries the repartimiento system of forced labor became, as elsewhere in Mexico, the dominant form of social relation between the Spanish/mestizo and indigenous populations of the isthmus.[24] In it, los indios were expected to deliver periodic tribute to the local Spanish alcalde mayor, a "creole functionary" position of the type described by Benedict Anderson.[25] According to historian Judith Francis Zeitlin, the alcaldes mayores of Tehuantepec typically demanded three repartimientos of manufactured clothing during their two years in office. This seems to have been tolerated by the Zapotecs; tributary relations to northern lords were familiar, dating to the late Aztec period. But when individual alcaldes and their indigenous cacique allies intensified their demands, violence often ensued.

The paradigmatic case was the Rebellion of Tehuantepec in 1660, when unusually harsh repartimiento demands for cotton cloth led to a protest by several thousand in Tehuantepec.[26] The Casas Reales were set ablaze; the alcalde mayor, don Juan de Avellán, was stoned to death along with a cacique ally; and the local indigenous authorities were removed from office.[27] For more than a year, Tehuantepec stayed in open rebellion against the crown until the forces of Oaxaca City and Mexico City were able to capture the leaders of the rebellion and subject them to a variety of brutal punishments.[28]

The violence surrounding repartimiento was real and dramatic. Yet historians argue that repartimiento was not simply a relation of naked oppression but was also imbued with a complex moral economy involving notions of acceptable tribute and patronage. At least some Spanish religious and secular authorities were also concerned about abuse of Native labor power. The istmeños in turn contested Spanish authority continuously throughout the repartimiento era and seemed to maintain a strong commitment to traditional political communalism. There were many cases of indigenous cabildos (councils) being overturned when it was felt that they had become too complicit with the Spanish.

As the indigenous population finally began to rebound in the early eighteenth century, demands for indigenous political and economic autonomy became stronger. Another major riot in Tehuantepec in 1715 led to increased anxiety about Native revolt, extending to the viceroy himself. Fearing also that the Natives might simply flee their pueblos to avoid further taxation, the viceroy allowed the indigenous political leaders of Tehuantepec more voice in selecting their local cabildo authorities.[29] As elsewhere in Mexico, the scarcity of labor and the fragility of its domination also lent Oaxaca's indigenous peoples some agency in their negotiations with the crown.[30]

Tehuantepec was indisputably the locus of Spanish-indigenous relations and conflict in the isthmus during the seventeenth and eighteenth centuries. Its status as a center of colonial administration, the presence of the only Spanish and mestizo enclaves of any size in the southern isthmus, and the close working relationship between indigenous and Spanish authorities eventually made Tehuantepec relatively state oriented and politically conservative relative to neighboring communities in the isthmus. By the time of Mexican independence, it was clear that Juchitán—a growing market hub benefiting from its proximity to the ranching haciendas, the salt flats, and abundant hunting and fishing resources as well as from longer distance trade from Chiapas and Central America—had become the more animated center of Zapotec autonomy, one increasingly known for its resistencia to external political and economic impositions. Although Benito Juárez would accuse the Juchitecos of "immorality and disorder" since "very ancient times," it seems more likely that Juchitán became the center of Zapotec autonomism and resistance only in the first half of the nineteenth century as Tehuantepec became acculturated to conservative political forces and as Juchitán's own size and regional economic power grew. European travelers' accounts throughout the nineteenth century uniformly depict Juchitán as a space of especially strong indio traditions, identity, and political unity. It is also not difficult to imagine that Juchitán had been for some time precisely the sort of place that istmeños might escape to if they were dissatisfied with the hegemony of the Spanish and their cacique allies in Tehuantepec.

The image of the guerrero Juchiteco was then honed in a series of revolts led by a prosperous Juchitecan rancher, Che Gorio Melendre (José Gregorio Meléndez to the Mexicans), who fought against the efforts of the Oaxacan state and the Mexican government to raise income by privatizing and monopolizing isthmian salt deposits. Beginning in 1834, and continuing episodically until 1853, Che Gorio and his supporters fought to maintain traditional communal rights to salt extraction.[31] He attacked the state military garrison at Tehuantepec in 1847 and declared the separation of the isthmus from the state of Oaxaca.[32] The uprising became perhaps the greatest challenge to the authority of Governor Benito Juárez's administration. Although Juárez sought conciliation, claiming that traditional communal use rights would not be disturbed by the new salt monopolies, neither his promises nor military action—including the possibly inadvertent burning of Juchitán in 1850—seemed able to quell Melendre's rebellion and guarantee istmeño acceptance of Oaxacan authority. State militia victories would see Melendre and his allies drift back into the forest only to reappear weeks or months later as guerilla fighters, burning haciendas and encouraging Juchitecos to "rob" the salt flats. In January 1853, the rebels again occupied Tehuantepec and contributed forces to the occupation of Oaxaca City in February. When in April of the same year Antonio López de Santa Anna returned for his eighth and final stint as Mexican president, he quickly signed a decree creating the

Isthmus of Tehuantepec as its own federal territory separate from Oaxaca and Veracruz. Melendre, though, was found mysteriously poisoned the very same morning. His dream of istmeño political independence was realized, but it only lasted two years, the length of Santa Anna's last term.

Viewed in the context of Mexican governmental efforts to license a rail corridor across the isthmus to foreign developers (as early as 1842) and the Gadsen Purchase Treaty of 1853—which established in the isthmus what amounted to a free trade and movement zone for US persons and property— Santa Anna's support for istmeño independence has to be viewed as something more than a capitulation of state authority to unruly indigeneity. Santa Anna and his successors (including Juárez) faced crippling foreign debts, uncertain incomes, and political unrest across the country. Transit rights across the isthmus seemed a safe bet to attract international interest and capital of a scale that dwarfed colonial repartimiento incomes and privatized salt and ranching schemes. Throughout the 1850s, until the McLane-Ocampo Treaty of 1859—which would have guaranteed the United States full use "in perpetuity" of the isthmus—failed to pass the US Senate, the promise of a railroad or a canal corridor across the isthmus brought a slew of northern explorers, surveyors, engineers, and investors to explore the possibility of transoceanic passage.[33] Among the antieólico resistance, we heard the canal plan frequently invoked as the first invasion of transnational capitalism into the isthmus and as the harbinger of all megaproyectos to come.

Historian Francie Chassen-López argues that the period from the 1830s to the 1850s was critical for the formation of a radicalized indigenous identity in the region: "The isthmian wars demonstrated how far the Zapotecs of Juchitán were willing to go in defense of their usos y costumbres, particularly access to land and natural resources. The Juchitecos, and their allies in other local ethnic groups, refused to accept defeat, and devised numerous forms of contestation (violence, land invasions, destruction of property, and armed rebellion), which forced the government and military leaders to negotiate with them."[34]

During the Reforma period (1854–76), the Juchitecans' reputation for violent independence bordering on savagery only grew as they became more directly and visibly involved in national political conflicts, such as the French intervention and then in the battles that led to Porfirio Díaz's dictatorship. In the struggle between Díaz and Juaréz for the Mexican presidency, Juchiteco guerreros captured, flayed, castrated, and killed Porfirio's brother, Felix, for having viciously attacked Juchitán while he was governor of Oaxaca. According to local legend, the particularly violent nature of Díaz's death was meant to repay his mutilation of the statue of Juchitecos' patron saint San Vicente as he sought to pacify the popular local Zapotec leader, Albino Jiménez, better known as "Binu Gada" (nine lives).

When Porifirio Díaz became governor of Oaxaca in 1881, this set off another wave of repression and resistance. Federal forces sought to bring Juchitán under control by fighting rebels, burning villages, and installing their own preferred political leader. But the next year a new Juchiteco rebel movement arose, killing the jefe politico and inciting Díaz to lead troops from Oaxaca City in 1882 to lay siege to Juchitán and kill or exile a large part of the male population. As Colby Ristow writes, "For the next thirty years Díaz maintained control in Juchitán by appointing handpicked *jefes políticos*, and transforming nearby San Gerónimo Ixtepec into 'one of the biggest garrison towns in Mexico.'"[35]

But the political tensions between the rojos (reds) and verdes (greens) in the last decades of the nineteenth century endured—with the reds identifying with more elite urban political, economic, and religious interests, including the Porfirista *científicos* (scientists),[36] and the greens identifying more closely with local ranching and campesino interests—simmering throughout the Porfiriato.[37] Another native Juchiteco leader, José "Che" Gómez, rose to local prominence, representing the verdes. As Díaz's regime collapsed in 1911, Gómez led another large rebellion against Oaxaca City.[38] Although the uprising cost Gómez his life, the uprising became a new "cornerstone of Juchiteco exceptionalism."[39] By this point, that exceptionalism was not solely a regional matter; Juchitán had come to occupy a very specific place in the Mexican national imagination as a specter of the violent indigeneity imagined to lurk beyond liberal-democratic reform.

Colby Ristow writes: "As the democratic-nationalist revolution threatened to open up the domain of the political in Mexico, agents of the center developed a discursive repertoire of marginalization that represented the people of Juchitán as particularly dangerous to civilization. Travel narratives, political addresses and newspapers represented the Juchitecos as insufficiently civilized to manage political life, despite their relative economic prosperity and active participation in the political conflicts of the period. Specifically, elite representations of Juchitecos portrayed them as pre-political, insular and violently contentious. . . . These representations mobilized recognizable symbols of barbarism—Indianness, unchecked women and the Orient—to underscore the unsuitability of the Juchitecos for inclusion in the political life of the nation."[40]

At the same time that Juchitán's political exclusion became a license for various forms of governmental violence and expropriation during the

Porfiriato, Ristow argues, Juchiteco exceptionalism also prompted a more significant incorporation of regional, indigenous interests into Mexican national and Oaxacan state politics. "The Juchitecos' insistence on the political and patriotic nature of their contentious politics challenged future revolutionary regimes to expand the national political sphere to include the poor and indigenous peoples of the periphery. As national power shifted during the revolution, Juchiteco demands for political incorporation found more open responses from the federal government, culminating with the regime of Lázaro Cárdenas."[41]

The pivotal figure in this process was General Heliodoro Charis (see chapter 2), the dominant political figure in Juchitán for much of the mid-twentieth century, an illiterate Juchiteco campesino who rose through the ranks to lead the first authorized battalion of istmeño soldiers in the Mexican National Army. A trusted ally of President Álvaro Obregón, Charis became an esteemed veteran commander in the postrevolutionary army, and during the 1920s his Thirteenth Battalion helped repress a number of uprisings, including indigenous uprisings, in other states. Despite pressure from the Mexican military hierarchy, Charis maintained Zapotec as the operational language of his unit and recruited only in the isthmus for new soldiers. Indeed, only the Yaquis supplied more indigenous soldiers to the Mexican army during this period, and Charis's was the only indigenous battalion willing and trusted to operate outside its native region. Benjamin T. Smith concludes, "By cultivating [a] clear-eyed ethos of sacrifice under-pinned by the expressed expectation of government reward, Charis and his men gradually forged an alternative to the threatening stereotype of the armed Indian: the obedient, loyal, self-sacrificing Juchiteco."[42] Political inclusion finally beckoned.

In exchange for his loyalty to the revolutionary nation, Charis was allowed to create a secure verde power base for himself in Juchitán that endured largely without conflict with Mexico City until his death in 1964.[43] In May 1930, Charis marched hundreds of his demobilized soldiers onto six thousand hectares southwest of Juchitán and established the Colonia Militar of Álvaro Obregón, named for Charis's federal patron. For the next thirty-five years, Álvaro materialized the power of Charis's network in Juchitán and the isthmus. Smith writes: "The armed men, stationed a few kilometers from the city of Juchitán, provided the local boss with the realistic threat of armed force if opposition politicians sought to take control of the district. . . . Some troops also doubled as extra-legal *pistoleros* (gunmen) ordered to assassinate or intimidate particularly recalcitrant enemies. Finally, the former members of the Thirteenth Battalion regularly provided Charis with a group of mobile and willing voters."[44]

Since 2012, the town has become a notable site of resistance to wind power development.[45] The town of Álvaro Obregón—capturing both the accommodation of revolutionary indigeneity to mestizaje nationalism and the perduring Juchiteco commitment to maintaining local Zapotec autonomy and power in the face of extranjero threats of every scale and type—epitomizes how much the complex and contentious history of the isthmus overdetermines contemporary support for and rejection of wind power development. The work of contemporary asambleas like the APPJ echoes across time, summoning the grinding repartimiento relations of the seventeenth and eighteenth centuries, the contested plans to privatize and cede istmeño land to foreign elites in the nineteenth century, the decades-long cycle of violent repression and rebellion experienced in the isthmus between the 1840s and the 1920s.

The lesson of this history is that Guidxiguie'—"city of flowers" in diidxazá—both does and does not belong to Mexico. Never fully assimilated to empire, Juchitán has long served as a powerful thorn in the side of Oaxacan and Mexican governance. It has represented the violent ungovernable foil against which Mexican settler liberalism has measured its capacity for justice. In the twentieth century, both during the revolution and again during the COCEI uprising of the 1970s—which many regard as the first major crack in the apparatus of the postrevolutionary PRI party-state—Juchitán has helped to give substance, fire, and purpose to opponents of Mexican mestizaje nationalism.[46] Even resisting the urge to mythologize the revolutionary spirit of Juchitán, it seems fair to characterize the city and the isthmus as a whole as an especially important site of political potentiality within Mexico.[47]

The coming of aeolian politics to Juchitán over the past decade has reignited these past potentialities and catalyzed what seem to be new political possibilities, a communitarianism that resonates as vibrantly with the global resurgence of neoanarchist, neoautonomist, and indigenist politics as it does with Zapotec traditions of political communalism.[48] For some, wind power will give Juchitán and its environs long-sought-after autochthonous economic power with which to match its political fervor. For others, the proyecto eólico threatens a final sacrifice of Juchiteco independence to the infrastructures of national and transnational governance and capital, a capitulation to el megaproyecto that has been more than two centuries in the making. The argument between these positions—with two different views of Juchiteco power vying with one another for prominence, inflected

FIGURE 5.6. Asamblea Popular del Pueblo Juchiteco

by experience, class, profession, gender—burns brightly in Juchitán today, echoing the red/green conflicts of the past.

La Séptima

This was true above all in la Séptima, for decades the epicenter of non-elite, indigenous politics in Juchitán. We found that arguments concerning the future were at their most sensitive and were raw in the barrios whose livelihood—fishing and field labor—seemed most at risk from the proliferation of wind turbines across the landscape. In these neighborhoods, despite increasing worries about "cultural homogenization,"[49] the language of the markets and street remains diidxazá, and the strength of binnizá identity and culture is not doubted. We go, for example, to meet with Melanie's friend, Na María Tecu, the one whose La Ventosan aunt had convinced her of the merits of wind power. But, just a few days later, we find that she has reversed her opinion, at least in speaking with a couple of extranjer@s.

Na María is cleaning fish in her courtyard with the help of her two daughters. Several relatives and neighbors poke their heads in to have a look because they have heard that there are gringos here. There is a shrill whis-

FIGURE 5.7. La Séptima

tling and cackling chorus of zanates in the trees that shade the courtyard,[50] almost loud enough to make human conversation difficult. The courtyard has been created by the cinderblock homes surrounding it, decorated with a number of crucifixes painted on the walls, dominated by three large fish-drying racks made of the naked rusted-metal frames and springs of mattresses. Na María continues working as we talk, sharpening a small dull knife on a whetstone, gutting the fish, cutting it in half from head to tail, and putting the fish on a drying rack. From time to time a zanate will circle down from the trees above to tear a hunk of flesh free before retreating swiftly back to the skies. A rooster pecks around below the cleaning station, looking for scraps. Later in the day Na María smokes the dried fish in a clay oven fired with charcoal and covered with thin pieces of metal.

Na María says that the whole Séptima is against the eólicos and, emphatically, that she has always been against them too. "We don't want any of that here. Nothing else will grow once they take the land. They will take everything! They will take the sea! They will create a sound that will scare away the fish!"

Melanie tells us later that she is surprised by how fast Na María has changed her mind; but charged information is circulating quickly through the barrio, inciting opinion first in one direction, then in the other. Melanie suspects shame as a partial motivator as well since Na María is known locally to have accepted a payment from Gas Natural Fenosa in exchange for her support.

FIGURE 5.8. Drying fish

We ask Na María cautiously about her aunt, the proponent of wind power from La Ventosa.

Perhaps suspecting our motives, Na María does not acknowledge having family there. But she does say, "In La Ventosa, there's no fishing, that's why they have the eólicos there. Here we are *puro pescador y campesino* [purely fisherfolk and campesinos]. That's why we are going to resist in this pueblo, none of us want these eólicos. What will our children eat if they ruin the fishing?"

We mention that there was a march in favor of the Playa Vicente project just a week before.

Again, she waves this off with conviction. "Just the *sindicatos* [union people], all the campesinos are against it. Even in La Ventosa it is just the *gente de dinero* [people with means] and the *maestros* [teachers/intellectuals] who want the eólicos. All the poor people are against it."

At a house down the street, we talk with Rosa and her sister Candida, who express much the same concerns about the eólicos. The people of the Séptima have been fisherfolk and campesinos as long as anyone can remember. The sea is the ultimate safety net, where one can always turn for sustenance when all other sources of work and food run out. Rosa and Candida are deeply concerned about putting their relationship to the sea at risk with noisy turbines; the loss of security is not worth whatever *dinerito* (little money) the landholders would pay them to stay quiet.

"They want to take our beach, they want to take the food from our children. We don't want the eólicos," Candida says. She says she has heard that the trees die when the eólicos are placed near them, they kill the coconuts and the mangoes by drying the land. And the *aeroes* (turbines) also make the water hot driving the fish away.

Rosa tells us that the mayor of Playa Vicente is giving money to many people to accept the park, money he has received in turn from the wind companies. She has heard that some people there support the park but not the majority. "Every party in power sells out to the eólicos."

But Candida's husband, Luís, who works on an oil platform off Campeche and who is enjoying a brief vacation at home, is not so sure about this resistance in the barrio. He worries about the lack of work in Juchitán, and he blames this on the líderes (political leaders) who scare away businesses. Those líderes, regardless of whether they are PRI or COCEI, "are dedicated only to themselves." The one invests in large ranches, the other in a transportation company, and another has purchased a hotel in Huatulco. "Probably more people would invest in Juchitán if they didn't worry that these políticos would steal their things."

Luís thinks that the fishing in Juchitán is slowly dying. He says that it still supplies the livelihood of thousands of people in the region, but over time their numbers have been declining. People are looking for other kinds of work, and they are traveling away to find it. He is one of those people and says that working on the platform is a good job; it pays well. He wishes there were more factories in the isthmus. Again, the líderes are responsible, having driven them away both south and north. "Look at how well even Chiapas is doing now!" So, in the end, the wind parks are "OK by him," although he laughs and acknowledges that he is not in the majority in the Séptima. Some people think the parks will cause radiation, he says, others fear the eólicos will heat the land or cause cancer. He's doubtful of those effects, but he knows fishing well and avers, "Noise contamination is a real thing! The fish will flee it, especially when the noise is constant, not just like a jet plane passing." The fish want nothing more to do with the eólicos than the campesinos do.

FIGURE 5.9. Café Sante Fé

Santa Fé

The outlook is very different at Café Santa Fé, which has been described to us as the place "where business gets done in Juchitán." These days the wind parks represent a big chunk of that business. The Santa Fé is an oddly unassuming roadside restaurant on the Panamerican Highway just west of where it crosses the highway to Ixtepec. Its artificial brick walls and bright-white tablecloths contrast the grit and scent and wind of much of the city, a world away from the convoys of truck trailers hauling goods, materials, and equipment down the highway and from the packs of gaunt stray dogs patrolling its shoulders. The food is simple but definitely not *rústico* and is served on quality tableware. The staff of the Santa Fé are professional and discreet, gliding in and out as needed, but giving a respectfully wide berth to the groups of men (and more rarely women) hunkered down for serious conversations, often lasting hours. The Sante Fé's back room can be cordoned off for larger, private meetings and occasionally becomes the site of press conferences. Its location is ideal for individuals transiting across the region. All this contributes to an atmosphere of an important crossroads where people can easily come and go, do their business, and remain strangely invisible in a transitory publicity. The Santa Fé's clientele is diverse, but there will always be at least one table, and often several, occupied by people working in or with the wind industry. Local politicians also often circulate through, pressing hands and slapping backs. If one wishes to see the politics of istmeño wind development in action, then one must visit the Santa Fé.

And we did spend countless hours there, observing quietly, chatting with friends, making new ones, hearing the latest rumors, doing impromptu interviews whenever we had the chance. Talk often focused on local political alliances and rivalries, where the latest blockades and land occupations were and why, how various wind park projects were faring, what new contracts might be in the offing. At times the rumors were bolder and more unsettling, like the one that the members of the powerful PRIista Gurrión family were laundering money for the Beltrán-Leyva cartel. Less frequently, one could also meet key figures among the antieólicos at the Santa Fé. Although the café's aesthetics did not suit them, like us, they found it a convenient meeting place and a valuable information hub.

In one encounter, we were asked to organize a fake interview with an APPJ member so that he could surreptitiously observe another table where, the asamblea suspected, the infamous PRI state congressman Francisco García López (a.k.a. Paco Pisa, see chapter 3) was planning to bribe two San Dionisio del Mar comuneros to buy their support for the Mareña project. Our friend in the APPJ wanted to make sure that our recording equipment was very visible so that "it doesn't look like I am taking money from gringos myself." The comuneros arrived along with a man whom our friend described as the broker who had set up the meeting. He hoped even to capture Paco Pisa on camera with the *bolsa* (bag of money) in his hand. For about twenty minutes we taped an interview in which our friend discussed the judicial amparo against the Mareña project and how little faith the inconformes had in receiving justice through any arm of the Mexican government. But the San Dionisians left, and Paco Pisa never showed; either the information was bad or "perhaps they changed the meeting place once they saw we were there." Such were the intrigues that could unfold at the Santa Fé.

For that very reason, the Santa Fé is haunted by local journalists keeping a finger on the pulse of local politics. There is always at least one press table there, sometimes two, with laptops out, cell phones spread across the table. Mostly they chat with one another as they write up their stories; sometimes they are lucky, and an itinerant politician or prominent citizen will offer impromptu interviews and statements. Our friend Rusvel from La Ventosa was often at the Santa Fé. On one occasion he introduced us to several of his colleagues who write on wind power often and thoughtfully, including David Henestrosa and Faustino Romo.[51]

David writes independently for a number of periodicals in the isthmus and publishes his own weekly paper. He studied engineering "but was a bad engineer," he says laughing. He does not consider himself a *periodista* (journalist) because of his lack of training, but he enjoys practicing journalism all the same. Members of his family are actively involved in the PAN party in Salina Cruz. The PAN is very weak in the isthmus compared to other parts of Mexico, he explains, but it gives him insight into how local power operates.

David is highly critical of how wind power development has taken place in the isthmus and only becomes more so over the months as we speak with him. He says the companies have largely been acting like "conquistadors." He is, however, a supporter of clean energy and would like to see Mexico's renewable energy transition advance, only in a more equitable way for all parties involved. He says he would even support the Mareña project "if they could guarantee real benefits to the communities whose lands they are occupying." He does not think forcing the companies to provide social development projects is the right strategy since "that is the responsibility of the government." What he would like to see instead is real profit sharing with the communities who are currently only getting a fraction of what they are due.

He says one of the most striking things about the eólicos is how they have brought the bitter rivals of the PRI and COCEI together in a common cause. "They are at each other's throats about almost everything else, but they come together to support the wind parks." David had more respect for "the Left" in Juchitán before the parks came. "COCEI has become *devaluada* [devalued] more and more; they are oriented by money rather than by leftist ideology. And the wind parks have brought the opposing factions of COCEI together like nothing else. The factions led by Héctor Sánchez and by Leopoldo de Gyves, Alberto Reyna, Mariano Santana, they are all involved in the parks. It's difficult these days to differentiate between pure PRIismo and the social líderes of the Left and their children: Emilio de Gyves, Lenin López Nelio, Pavel López. They respond only to political and monetary interests; they have no true connection to leftist ideology."

We ask him how much journalists have themselves become involved in promoting or criticizing the wind parks.

"It's very hard to be independent in Mexico," he says and shakes his head. "The local media need to eat too, and the biggest papers in the region, *El Sur* and *El Sol del Istmo*, are controlled by the Gurrión family. There is a lot of coverage that is bought here either directly by the wind companies or by the interests that support them." David sees himself as one of the few journalists willing to take seriously the inconformes' resistance and their concerns about wind development. "With some of them, like UCIZONI, you sense an ulterior motive, a desire to create trouble and get attention for themselves. But then you meet people like Mariano [López Gómez of the APPJ], and he's such an honest guy, a good kid; he's from a humble background and just is very passionate about his politics."

David reiterates this opinion again, when, in late March and early April 2013, the political situation in Juchitán turns very ugly. The dual blockades of the Mareña and Gas Natural Fenosa construction sites are countered by a wave of violent intervention from the park's supporters. Radio Totopo's broadcast equipment is stolen, and Beedxe's arm is broken. Mariano is detained by the police on the suspicion of having demanded ransom money for construction equipment taken by the APPJ. Anonymous threatening phone calls and less anonymous visits from *grupos de choque* (groups of thugs) cause some leading APIDIIT and APPJ members to go into hiding.

David says he thinks the APPJ has a just cause and is all the more sure after receiving a veiled threat at a traffic stop by state police, saying that he should consider "toning down" his coverage of what is happening in Álvaro. He sends us a message on Facebook, asking us to take note in case something happens.

We ask whether he thinks all this is happening because the supporters are getting scared.

He replies, "The people who play politics around here are probably used to threats. But for me, this is something new."

Faustino meanwhile works for the Noticias network, a news organization with a historical reputation of being anti-PRI and a current reputation of being pro-Cué. Faustino seems neither, more a patient analyst of the complexities of istmeño politics than an advocate himself. He softly criticizes David for being "too one-sided" in his coverage of the wind parks. "He seems to favor the resistance. I think you always need to see both sides of it."

For Faustino, that means reckoning with the real benefits these parks might bring. "As I see it, wind power has a global benefit, and Mexico is a strategic site for the development of this potential. And Oaxaca, particularly the isthmus, is a strategic site within Mexico because of its resources. So far, so good. But what benefit will it bring to the communities upon whose land the parks will be built? They talk about a $1.2 billion investment. That could be very significant, something important for this region that hasn't received those magnitudes of investment until now. But, OK, then you also have to ask, 'Investment for whom? What part of those resources might directly or indirectly benefit people here? And what is going to the turbine manufacturers, the cable makers? What part of it will end up staying in Oaxaca [City]?' They never break that information down."

He winces at the attitudes of some wind developers. "When we asked the representative of [the Mexican wind energy lobby organization] AMDEE

FIGURE 5.10. Land invasions, Juchitán

why the projects did not incorporate community-wide benefits, he gave a nasty laugh and said, 'Well, when they invest in and build their own proyecto eólico, they can do what they like with the profits.'" At the same time, he sees the opposition being steered by disreputable political groups like UCIZONI and the blockade on the road to Playa Vicente being run "by gangs of delinquents who just like to go out and fight with the police."

In his journalism, he becomes increasingly critical of the inconformes during the time we are in Mexico, and it is not clear to us whether pressure is being exerted on him to do so or whether he, too, has just become frustrated by the retardation of economic progress in Juchitán. Faustino is no campesino, rather a Spanish-speaking middle-class professional who is not obviously committed to any ideological agenda. He talks to us about the land invasions that have been accelerating as another heated mayoral campaign between the COCEI and PRI looms. The rival factions of the political parties are positioning themselves for advantage in the primary elections and establishing "vote banks," much as Heliodoro Charis once did in Álvaro Obregón.

In late April he tells us, "Last week alone there were six new land invasions. The municipio estimates there are twenty-five invasions total going on now. That's a lot of land! In the old days, the invasions were about getting homes for people with no homes. But now most of the invaders already have homes. They're doing this professionally for one faction or another. It's about securing votes, bribes, extortion."

Faustino greets all this with a mixture of resignation and wry humor; he is also fond of jokes and pranks. When the alleged Mayan "end of the world" prophecy fails to materialize in December 2012, he issues a fake news bulletin via his Facebook page: "Breaking news: the Mayans have announced that nothing will happen in the isthmus because COCEI, Sección 22, the mototaxistas, the taxistas, PRD, PRI, PT, the construction workers, the campesinos, the sorghum farmers, all of the above are blockading the highways, conspiring to prevent the end of the world."

How Caciques Think

Our journalist friends like Faustino, David, and Rusvel did not exactly say that the líderes would lie to us, but they did not seem hopeful that we would learn any truths from them either. So much of the politics of Juchitán takes place at the street level—in the barrios, in the occupied lands—where different networks operate their own fleets of mototaxis and vie for territory and alliances. No líder would admit to a gringo what it really takes to maintain power in Juchitán. Still, it would not hurt to ask. You might glean insights into the view from liderazgo as to how a cacique might think about things.

Our efforts to make contact with local Juchiteco PRI and COCEI leaders bear little fruit overall. We acquire their phone numbers easily enough, but cold calls prove ineffective, and occasional promises of personal contacts do not pan out. The líderes never seem to be where they are supposed to be. But we have met Daniel Gurrión—the mayor of Juchitán, scion of the Gurrión construction empire—at the FIER event in Oaxaca City (see chapter 3), and he has offered to meet with us on our next visit to the isthmus.

Gurrión and his entourage (a burly driver and two underlings who say not a word through our entire trip) come to collect us in a brand-new white Chevrolet Suburban. Since the winds are high, he suggests we drive out to see one of the wind parks near La Ventosa. Gurrión is nothing if not a charmer, wielding power in a velvet glove. He establishes his elite, cosmopolitan credentials quickly by sprinkling a little English into his discourse, telling us about his former life as a dentist, his vacations in Europe and the United States. He also has, thankfully, the nothing-at-stake candor of an oligarch.

As the car winds its way through the heart of Juchitán, Gurrión gestures through the wind toward the crowds of vendors, rewriting the indigenous essence of the city not as warrior but as trader.

"Basically, our ethnic group, the isthmus Zapotecs, is a group characterized by its commercial activity. We have been *comerciantes desde siempre* [merchants since forever]." Although he recognizes that there is still much poverty in Juchitán, he attributes what wealth the Zapotecs have to their commercial resourcefulness.

His crown weighs heavy at the moment. He speaks at length about how difficult it is to manage Juchitán. "It's hard." Gurrión makes an airy motion with his hands. "We don't have recreation areas for children or parks for sports. And we have a lot of trouble with alcoholism and drug addiction here. That is true elsewhere in the country, but I believe the statistics say that our consumption of beer is among the highest in the country."

We pass several stands of mototaxistas, whom we know count among the most predictably unruly political presence in the city; they are foot soldiers for various networks, some licensed, some unlicensed, skirmishing also with the normal taxi drivers for the legitimacy to move passengers across Juchitán's streetscape. Gurrión gestures to them as well.

"The state government made a bad decision recently to give out two thousand new permits for the mototaxistas. It's made things even more difficult here in terms of the traffic, and they constantly fight with one another."

By the time we are on the highway heading east toward La Ventosa, the conversation has shifted toward COCEI. In the beginning, Gurrión says, the movement was united and *muy bonito, muy bueno*. He even participated a little in its fringes when he was young. Then he picks up a piece of paper and begins to draw what looks like a pizza, a circle with wedges filled with dots.

"Over time," he says, "it began to be contaminated, distorted; it subdivided into various *líderes*, and each *líder* had his own group, his own little group, and taken together, they represent a huge problem here now. Every one of them wants to maintain strength, to maintain people, and what they do is they seize some plots of land, they take them, invade them, and they found *colonias* [colonies] and give them to their followers who then vote for them in the elections. It's their modus vivendi: they blackmail people, block the highway . . ." He trails off, not frustrated, just lost in his thoughts.

"You know, maybe two or three days ago, I met with the people from Wal-Mart, they were interested in investing here, but I don't know whether it will go ahead. With the politics, someone would blockade the construction site, blackmail them, if the store opened, more *chantaje político*."

He speaks at length about *chantaje* (blackmail), how it has become an industry itself, the spirit of comercio in predatory lawless form. "Yes, well, Juchitán *es muy bonito, pero es muy complejo* [it's very nice, but it's very complicated]."

We stop at an overpass overlooking a wind park, but the winds are so strong that we can scarcely speak to one another. We are blown back and forth by gusts as Gurrión gestures at the expanse of the turbines, which stretch far into the distance; he is telling us some stories about how the park came into being, but we cannot even hear him. A massive Grupo Bimbo tractor trailer whooshes past, more proof of Juchitán's robust commercial life. On the ride back, in the mobile quiet of the Suburban, Gurrión talks about the proyecto eólico, and he shares again the concerns he articulated in Oaxaca City. The federal authorities do not seem to be regulating the industry effectively, or rather they "regulate them in an arbitrary manner" and some projects—he mentions Acciona again specifically—seem to have been allowed to avoid paying their legally required cambio del uso del suelo (change of land use) fees to the municipal government.

"Here in our downtown market, there isn't a single vendor, even the ones who are practically sitting in the street with a tiny stand, the nopales vendors, the egg vendors, every single one of them pays their city taxes. So why not these wind companies?"

He remains troubled by the elusive benefits of the parks as well. The question of equitable exchange persists. Why should the commitment of land to electricity not be rewarded with cheaper electricity?

"The people who live in this region had great expectations about these parks and the benefits they would bring. And I am not saying there is no benefit because there obviously is a benefit. But, honestly, it hasn't lived up to the expectations. The electricity is still very expensive even though we are now creating all the electricity for the region."

We ask him about the Ixtepec model of community-owned wind power, whether that is something he could imagine supporting in Juchitán.

He says it would be ideal, but he wonders whether it would be legally and economically viable. He would be happy, he reiterates, with just a few turbines to call Juchitán's own. "Just two or three, that would be enough, to provide the municipality electricity, to offset the bills from CFE."

By now we are having coffee at—where else?—the Santa Fé. Gurrión is in his element, leaving us to circulate from table to table, at home with the comerciantes.

In April, we are also able to meet with the COCEI candidate for mayor, Saúl Vicente Vázquez, who will eventually become Gurrión's successor when he wins the 2013 election. Vázquez has a strong background in human rights and indigenous rights issues; he is even a member of the United Nations Permanent Forum on Indigenous Issues and enjoys considerable support from Juchitán's binnizá intelligentsia as well as from Héctor Sánchez López's faction of the COCEI. His campaign office, however, is very humble, just a table and a few chairs and a single piece of white paper taped to the wall outlining strategy. Our meeting is compact and formal, and we get the sense that Vázquez is actively trying to avoid committing himself to any particular position regarding wind development, saying curtly, "It means a lot and there are different opinions."[52]

He says he has not seen any of the wind park contracts personally, but those who have seen them tell him they are quite unbalanced in favor of the companies. He also says that he believes the companies did not inform people fairly and sufficiently in advance of signing the contracts and that there has been corruption at all levels of government, with bribes given to purchase consent. On the other hand, he sees nothing wrong with the parks in principle; from an environmental standpoint they are an improvement over how Mexico currently gets its electricity.

"With all this criticism of the eólicos, I wonder why you never hear people protesting Pemex with the same vigor." He also notes with a slight smile that people have a habit of complaining about forms of development whose products they happily use. "Like mining, no one likes it. But then look at this fork. Its metal came from some kind of mining somewhere." Still, he says, wind developers should ultimately bear the responsibility for the protests. "It is their failures that have allowed doubts to grow into opposition."

Vázquez will not concede that the eólicos are a campaign issue, however. Shrugging, he says, "It would depend on whether that person felt the municipio could address his feelings about the parks in some way."

Perhaps Vázquez is being evasive or honestly reporting the herculean effort required to unite COCEI's many factions on any single issue.

This is the fact he wants to impress upon us most strongly: "One can't really speak of *the* COCEI. It doesn't exist in that way. It has never been a united political party. It's a social movement composed of many different autonomous organizations that coordinate with each other. And it includes the guy down the street who is COCEI but doesn't belong to any group. At last count, there are forty-eight different organizations involved in COCEI; all the ex-mayors have their own, and then are many more besides. So, with a wind park, for example, it could be that one part of COCEI is for it and another against it. That's how it is with us."

The inconformes are many and multiple in their backgrounds too, ranging from workers, campesinos, and fisherfolk to maestros and other intellectuals who have been drawn to the cause. The range of actors mirrors the elements of the COCEI movement in the 1970s, which fought street battles against political and economic PRIismo for much of the decade before finally taking control of the city government in 1981, and then continued to fight against repression from federal PRIismo for the next decade and a half. In the isthmus, as noted above, the PRI had only a relatively weak network until the mid-1960s, which left it susceptible to contentious politics. But while the inconformes who have organized against wind megaprojects today may identify themselves with the COCEIistas of the 1970s and 1980s and their struggle, they utterly reject what COCEI has become today. Their assessments are indeed more savage than those we heard from a blood-enemy PRIista like Daniel Gurrión. When we attend the asambleas and marches organized by the resistance, we rarely hear a sentence about the COCEI uttered without a snarling reference to the movement's degeneration and betrayal of the common Juchiteco. Leading COCEIistas' support for the wind parks, for the transfer of land to extranjeros, seems to have been something of a last straw culminating two decades of abuse, lies, and manipulation.

This is the perspective shared with us by Bettina Cruz Velázquez, one of the cofounders of the Asamblea de Pueblos Indígenas del Istmo en Defensa de la Tierra y el Territorio (APIIDTT) and quite possibly the most famous binnizá activist in the world today. She is not only lionized and vilified by turns in the Mexican national press but has also received substantial attention from international human rights organizations. Cruz has travelled widely in North America and Europe, including once testifying before the European Parliament, speaking against megaproyectos and for binnizá rights in the isthmus.[53] But Cruz began her activist life as a teenager in Juchitán within COCEI; members of her family are still deeply involved in the movement. She scowls when she talks about COCEI today, its leaders "seduced by money and power," a caricature of its former self.

Cruz's husband, Rodrigo Peñaloza, is another APIIDTT cofounder, and he sees in COCEI a cautionary lesson for all "vanguard movements," saying, "Any movement carries within itself the germ of its institutionalization, and for COCEI, that germ was Stalinism."

The purpose of APIIDTT, according to both Cruz and Peñaloza, has been to help nurture a different kind of political movement in the isthmus, one that is decentralized and antihierarchical and more strongly rooted in indigenous traditions of self-governance. "At first we were calling ourselves a *frente* [front] but that was too vanguardist; we realized that the better model was the asamblea [assembly]."

We have lunch with Cruz and her daughter, who is studying anthropology herself, early on in our fieldwork. This is before the violence around the Mareña and Gas Fenosa projects erupts and Cruz is forced to go into hiding with her family after receiving a number of death threats she deemed credible. We spend the afternoon in the Donaji restaurant, next door to the APIIDTT office. The office is itself remarkable for having been the last functioning office of Juchitán's bienes comunales before the COCEI leader and last president of the comuna, Victor Pineda Henestrosa (Victor "Yodo"), was kidnapped on July 11, 1978, by a group of men dressed in military uniform and disappeared. The removal of the comuna leadership paved the way for the expansion of private property in the region.

Cruz and Peñaloza view this as a form of land occupation itself, one in which the wind parks have been directly implicated. "All of the contracts that have been written to lease what was once communal land for wind parks are illegal. You cannot write a private contract over communal land." Rebuilding the leadership of the bienes comunales is one of the most controversial campaigns the APIIDTT is working for; another is their campaign against CFE's electricity tariffs.

Cruz believes that the APIIDTT is as close to a functioning asamblea as Juchitán has: "We began here in Juchitán, and here anyone is free to join and become incorporated in the asamblea. Every pueblo, every community, every committee, has its own rules and dynamics, though. What we do here is distribute information, research legal questions, deal with the media, engaging issues which benefit all the members of the asamblea. Our ultimate goal is to stop el proyecto eólico in our communities and on our land."

Cruz is disturbed by how the eólicos have contributed to the emptying of the primary economy of agriculture, driving campesin@s off their land to search for scarce work in the cities or abroad. "Working the land—farming, fishing, producing—is a way of life that has been abandoned and degraded in this country. But it is the heart of everything."

She is dubious of the wind companies not only as beacons of a white-collar future for the region but also as to whether they are in fact realizing positive change at a global level. "The same companies that are creating supposedly renewable energy here are producing thermoelectric power plants in other places. So what is this change they are talking about? I don't see change here other than in the discourse, and they are utilizing global sentiment [*sentimiento mundial*] to further their exploitation. What they are doing is monetizing climate change."

And she believes that the land, once taken for the wind parks, will belong to these companies forever. "When will they return the land? These are usufructuary contracts for thirty years or more. This will be their land forever; it will be ruined anyway for agriculture. So where is the advantage for the people here? There is no advantage. It is only for the government functionaries, for the caciques who have massive landholdings. The campesino is losing his way of life."

It turned out that the APIIDTT would win its greatest victory during the period of our field research when a federal judge upheld the amparo complaint of the inconformes in San Dionisio del Mar to block the Mareña Renovables project pending an investigation, an act that contributed eventually to the failure and relocation of the project. Meanwhile, getting Cruz's blessing proved invaluable for our research; connecting us to other asambleas comunitarias and asambleas populares across the region. Over the course of 2012 and 2013, with her help, we got to know key figures in the assemblies of Álvaro Obregón, Juchitán, and San Dionisio del Mar well.

Alejandro López López, a cofounder of the asamblea comunitaria of Álvaro, says he has known Rodrigo and Bettina since secondary school. They had all been doing different things, but the megaproyectos brought them together in defense of natural resources, especially land, "which is nature's mother, what connects the present to the future. Land is part of the family, it has a special meaning for the campesinos." Alejandro is not a campesino, but rather a maestro, and is very active in CNTE 22 in Oaxaca City. He helps the inconformes to draw upon the resources of the teachers' union to support their cause, which he says is not only to prevent wind parks from destroying the land and livelihood around Álvaro, it is also to wrest Álvaro out from under the control of the líderes altogether.

The asamblea's blockade of the access road to the Mareña Renovables wind park in November 2012 evolved over the next year into a challenge to the structure of political authority that had blanketed the town since the days of Charis. In August 2013 the asamblea voted to institute usos y costumbres law and to prevent the participation of political parties in municipal elections. Voting booths were burned, which in turn intensified acts of violence against the asamblea by its political opponents, and new municipal authorities from Juchitán were imposed in August 2014. The conflict remains unresolved today, but the asamblea comunitaria is, its members say, resolved to the defense and autonomy of Álvaro against all political parties.[54] The asamblea has organized its own community police force and its own community radio, both named for General Charis, who remains a powerful symbol of binnizá futurity and autonomy.

Alejandro foresaw much of this conflict when we interviewed him formally in early January 2013. "The betrayal and selling out of Álvaro Obregón has been going on for decades. COCEI has taken the force of Álvaro Obregón and treated it like a gold mine because Álvaro Obregón has always been *muy valiente* (very brave). Álvaro Obregón are a people of few words, but a decisive people, and decisive with their machetes. That is why all the leaders of COCEI—Santana, Héctor, Leopoldo—they are all coming at us like rabid dogs because we are depriving them of their gold mine."

He explains, "They used to split the government of Álvaro Obregón like a cake; Santana would pick the president, Héctor the treasurer, and so on. But that is all over now. We are done with selling out to the líderes. They are all rich, and it doesn't matter what party they represent, whether the PRI or the PRD or the PT [labor party] or the PAN. They have just worked to the divide the pueblo for their own gain. We are retaking Álvaro Obregón and looking for unity. That's why we call ourselves an asamblea, not a front, not a group."

The struggle against the líderes and the struggle against the eólicos merge together in Alejandro's discourse. The heart of the resistance to both forces, he says, is the sacred triangle of San Mateo, San Dionisio, and Álvaro Obregón because that is where the asambleas are still held in the indigenous language, where even Spanish is resisted. "It's all about history, about origins," he says. These asambleas are working together to spread their fuerza (force) across the region, to fight against the transnational companies and their political allies.

In several visits to Álvaro, we spend a good deal of time talking to members of the assembly, and the vast majority are not maestr@s but rather fisherfolk and campesin@s, and there are many women among them. Some are motivated by an ideological commitment to the politics of land and sea, some wish simply to safeguard their livelihood, but all remember the day when the company began its vegetation removal work on the nearby sandbar and began asking fishermen for ID cards to gain access to the place where they have put in their boats for as long as anyone can remember.[55] That was the moment, they told us, when the community realized something was going wrong, that this project was threatening to take their land and sea away from them. Their anger was still palpable months later.

"We are the people of Charis!" one man said. "If they want to see blood in the sand, let them come. We are ready."

FIGURE 5.11. *"Somos viento"*

In the summer of 2013 we begin to see the slogan *"Somos Viento"* (we are the wind) spray-painted across Juchitán. The idea of the resistance occupying the wind is growing.

Mariano López Gómez, one of key organizers of the APPJ, posts on his Facebook page, *"Somos viento y juntos seremos huracán."* (We are the wind, and together we will become a hurricane.)

Carlos Sánchez, the jaguar, founder of Radio Totopo, embraces the wind too, helping to cofound a new collective, *Binisá,* "which means the wind from the south, the soft wind that brings rain. But sometimes it turns into a strong wind, a wind that only begins slowly and then comes with force and destroys what lies in its path. Here what we will destroy is the silence, the silence that has fallen over the people of Mexico."

We also meet with the youngest generation of Juchiteco activists, the #YoSoy132 kids. Even though they speak Spanish, organize through social media, and offer detailed critiques of global neoliberalism, they continue to speak the language of an istmeño history of resistance.

Dalí Martínez explains to us, "The *lucha* [struggle] has been going on in the isthmus since before the time of the Mexican Revolution. There has been wave after wave of rebellion over the years. This is a tradition that we are proud of, and we want to carry it forward to the next generation. It is part of our identity, this history of lucha, so we want to defend it. When we heard about the eólicos, these transnational projects with negative social and environmental consequences, we became worried about them and wanted to fight them. We want to achieve the autonomy of the isthmus, that is the ultimate goal."

When we ask Dalí and his friends whether Occupy and the Arab Spring influenced them in any way, they say no; this is really about their cultural tradition of rebellion reinventing itself for each generation across time. This is about *los tecos valientes* (the brave Juchitecos) "the way it always has been."

The ways things have been in Juchitán will undoubtedly influence the future of aeolian politics in the Isthmus of Tehuantepec. But things have not only been one way; both the red politics of urban elites, landowners, and the commercial class and the green politics of campesin@s, fisherfolk, and migrants have shaped the history of the region in equal parts. Both greens and reds have been and continue to be further fractured into dozens of political networks defined by kinship, liderazgo, charisma, and ideology. Sometimes these networks affiliate with existing political parties or aspire to create new ones. Often they claim land, whether a mototaxista stand or a new colonia occupation on the outskirts of town. The anthropolitical terrain of Juchitán is immensely complex, a subject demanding detailed political ethnography and political geography far beyond the scope of the present volume. But in the zone where aeolian politics and anthropolitics intersect, we have seen how wind development has been avidly embraced by some as a means of concentrating wealth and power in the constant game of positional advantage in the city. In some sense, wind parks are the ultimate form of land occupation, land claimed by concrete and steel and fencing, bringing income to buy votes, trucks, new homes, and so much else besides. For others, meanwhile, we have seen how wind parks are excoriated as worst kind of megaproyecto development, the sinister collaboration of local caciques and transnational capitalists to complete a centuries-long project of capturing and expropriating the wealth of the isthmus. At the same time, even wind power's worst critics would acknowledge that the proliferation of wind turbines across the lands around Juchitán has galvanized political opposition and catalyzed a recommitment to asambleas and usos y costumbres the likes of which have been unknown for decades.

The reason to end this journey across Mexico in Guidxiguie'/Juchitán is that it epitomizes the analytic limits of universalizing power concepts like "capital," "biopower," and "energopower" for defining the political terroir of istmeño wind. Yes, these concepts have considerable value. Those who push private property, factories, wage labor, and consumerism in the region clearly desire to constitute the frontier ecology of alienated land, labor, and value that "capital" denotes so well.[56] "Biopower" is seemingly more remote to the everyday politics of the city, but when a political leader like Gurrión talks about cheap electricity and turbines dedicated to municipal energy supply, he is not just thinking of his budgetary bottom line. Although deeply and fiercely contested, the biopolitical aspirations of "development"—good jobs, good health, prosperity, growth—hang in the air both in the city hall and in the Séptima. "Energopower," too, can offer insight into istmeño aeolian politics, both in electricity's capacitation of biopolitical projects and in the infrastructures—grid, pipelines, highways—that crisscross the city and enable its many projects of modernizing and indigenizing. My point is simply that these concepts are minima not maxima. The aeolian politics of the isthmus are powerfully, *powerfully* shaped by identities and institutions that are not reducible to either instantiations of or reactions to capital, biopolitics, and energopolitics. Caciquismo and liderazgo belong to this spectrum, as do asambleas, usos y costumbres, and indigenismo, as do ways of life and forms of knowledge replete with soil and sea.

If we wish to imagine and discuss aeolian futures in the Isthmus of Tehuantepec—or, for that matter, anywhere else in the world—what is needed are more and better ethnographies of the multiplicity of forces (historical and contemporary, material and anthropolitical) that are generating, inflecting, and obstructing potential futures. It never ceased to amaze us how little the proponents of "Oaxacan wind power" in places like New York, Mexico City, and Oaxaca City seemed or cared to understand the reasons for its contentious politics in the isthmus itself. The depth and seriousness of contention meanwhile made perfect sense to anyone who had spent a few weeks talking to people in Juchitán. Part of building a future that does not endlessly repeat the Anthropocene trajectory is caring about forms of enablement that exceed those with which we are familiar, disabling our engines of epistemic and political universalization and rebalancing our analytical attentions and worldly engagements in favor of what is meaningful and valuable in the localities where the wind blows. How we can make new and better aeolian futures is the subject of our conclusion.

Joint Conclusion to *Wind and Power in the Anthropocene*

CYMENE HOWE AND DOMINIC BOYER

Aeolian Politics, Aeolian Futures

We went to the Isthmus of Tehuantepec as ardent supporters of renewable energy transition, and we left with that conviction intact. Wind power (alongside solar power, tidal power, geothermal power, and biofuels) has an important role to play in reducing greenhouse gas emissions and decarbonizing electricity generation. But we also returned to the United States with a more skeptical view of renewable energy's capacity to guarantee salvation from climate change let alone the Anthropocene. Renewable energy has a necessary but insufficient role to play in a process that will amount to a refashioning of the civilization(s) that brought us to our present ecological and political conditions. What our field research on Mexico's aeolian politics and the ecosystemic limits of wind power taught us above all is that it is all too easy for renewable energy development to occur with little or no social, political, or economic transition attached to it. It is both possible and common to build wind parks firmly within a model of resource extraction that is typical of global fossil fuel and mining industries. We have offered extensive documentation of such wind development in our Mareña (*Ecologics*) and La Ventosa (*Energopolitics*) case studies—where attempts to capture the wind resulted in failures, both human and other than human. We have also shown in the case of Ixtepec (*Energopolitics*) that other development models exist, even if they are being actively resisted in Mexico. Where human desires for energy are not in balance with their ecosystemic context, as we see across the *Ecologics* volume, there is little hope of remediating climates either locally or globally.

Taken together, the three case studies we have followed in *Energopolitics* and *Ecologics* demonstrate the turbulence surrounding renewable energy as the world awakens to the Anthropocene. They tell stories that are specific to Mexico and yet also exceed national boundaries. Carbon politics, finance capital, global industry, consumerism, and a lack of environmental protections have laid deep infrastructural grooves and have largely drawn aeolian politics into their orbits. Thus, the win-win-win visions of green financiers, entrepreneurs, and developers who promise that climate change can be reversed while maintaining everything else about the modern world, especially economic growth and a positive return on investment to shareholders, show a stubborn reluctance to abandon the structural deficits of carbon-based modernity. Those imaginaries are shared to a great extent by Mexican and Oaxacan politicians and technocrats who, steeped in neoliberal certainties and petropolitical anxieties, yearn for foreign direct investment to extend and improve the biopolitical functions of governance in the form of health, security, and prosperity. Some even believe that wind power can help to fulfill delayed or abandoned plans to bring, at long last, the isthmus into the nation, not as a repartimiento vassal but as vigorous organ of the mestizaje national body. Local leaders and asambleas, elected and unelected, are likewise drawn toward the biggest influx of international attention and activity the isthmus has experienced since the mid-nineteenth century. Some fight for local or indigenous autonomy and sovereignty against the encroachment of megaproyectos, others pursue windblown wealth to further dreams of better jobs for their children or the accumulation of capital and leverage or for the opportunity to extend and deepen their networks of influence. It is not only in Mexico that dreams of aeolian futures are paradoxical; what are heavenly images for some are nightmares for others.

This is only to speak of the anthropolitical dimension of aeolian politics. We must also consider the Anthropocene trajectories of birds and bats and fish, the machinic life of turbines, the grid, and trucks, the unruly howl of el norte, and the gentle breezes of binisá. Aeolian politics is always already more than human even if the ecological interdependency of human and nonhuman potentials is largely ignored in standard treatments of wind power. It is for this reason that we have created a duograph to offer not only an ethnographic division of labor in its coverage of the three studies but also an analytic division of labor that allows us to pursue, with better depth and peripheral vision, both the mapping of anthropolitical enablement and the mesh of human-nonhuman relationality that is often allowed to drift into the background of reckoning with the Anthropocene. Questions of wind and power circle each other in the Isthmus of Tehuantepec—How can the fierce northern winds be harnessed? With what machines? To what end? Benefitting whom? Displacing whom? Earning what? Killing what? For how long? And with what consequences? We have likewise sought to let the analytics of wind and power speak to each other in this duograph, probing their potential to remake and unmake the Anthropocene. Enablement is always relational: some complex of forces, things, and events begetting others. Relations, for the same reason, always enable. The riddle of the Anthropocene is what mesh of relations and actions will allow us to disable the reproduction of the present while being present in the production of a future. For those who wish to solve that riddle, we must attend to both human politics and all the other relations and forces that make those politics possible.

An earlier version of our duograph was titled *Winds of Desire* because everywhere we turned in Mexico, we found people wishing for the wind to deliver something: money, electricity, influence, legitimacy, prosperity, development, power. At times, desire cloaked itself in mathematics, rationality, and common sense. At other times, it reveled in naked hallucination. Those who desired were rarely satisfied with what the wind had already delivered to them. What desire always accomplishes best is the propagation of more desire. Here, at the end of a project that has been nearly a decade in the making, we are asking ourselves what it is that we wish from wind power. It turns out that our object of desire is also elusive and receding. Still, we are drawn toward it: we want better aeolian politics oriented toward achieving better aeolian futures.

Our final report to the National Science Foundation listed the following findings and recommendations based on our research:

> The field research for NSF #1127246 yielded several important findings and recommendations that will contribute to more positive development outcomes in Mexican energy transition in the future. (1) The dominant development model prioritizes the interests of international investors and developers and local Isthmus political elites over other stakeholder groups, especially the regional government and non-elite Isthmus residents. (2) The dominant development model has reinforced hierarchy and inequality in Isthmus communities through unequal distribution of new resources like land-rents. (3) The development model has generated significant polarization in Isthmus communities regarding wind parks and undermined trust in government and industry. (4) The financial benefits from land rents

are currently primarily being directed toward luxury consumption by elites. (5) A majority of Isthmus residents appear to favor wind power development were its financial benefits to be more equally distributed. (6) Project findings suggest that the Mexican government needs to re-evaluate its development model to guarantee (a) that entire communities and not simply elites are involved in project design and implementation, (b) that mechanisms be developed to guarantee that wind power development yields consistent and significant public benefits, and (c) that regional governments receive sufficient federal funds to develop a regulatory agency with the authority to guarantee that wind power development is truly transparent and beneficial to all stakeholder groups.

To put this in less muted terms, in our view, there will be no "renewable energy transition" worth having without a more holistic reimagination of relations in which we avoid simply greening the predatory and accumulative enterprises of modern statecraft and capitalism. In this respect, the record of Mexican wind development thus far does not inspire much confidence. The model of wind development that currently dominates the isthmus has been very effective at building wind parks, but it has done almost nothing to disrupt the toxic kinds of relatedness that made it necessary to build wind parks in the first place. It has left wind power in the thrall of finance capital, state biopolitics, and energopolitics; parastatal utilities and infrastructure; PRIismo, caciquismo, consumerism, and many other -isms besides. The case of Mareña Renovables (in *Ecologics*) came to absorb and reflect all these conditions and in so doing was stalled out of existence. In failing to account for local histories and imagined futures, and in repudiating local worries about environmental harm, Mareña's potential to provide climatological remediation and reduce greenhouse gas emissions was drowned among the fish. With the Yansa-Ixtepec project (in *Energopolitics*), we do find a scrappy DIY prototype for a better aeolian future, one that seeks to harness wind-generated electricity to help a rural farming collective to better guarantee their own autonomy and futurity while still contributing to the global cause of decarbonization. Yansa-Ixtepec has flaws to be sure—its benefits will not extend far beyond the collective, and it requires a grid and a failing parastatal electrical utility to pay its rents—but if the project is ultimately thwarted, Mexico will miss its best chance to connect the heady ambition to be a global leader in clean energy development with the interests, hopes, and worldviews of people living in places where the wind is strongest. In the end, we need not just new energy sources to unmake the Anthropocene, we need to put those new energy sources in the service of creating politics and ecologics that do not repeat the expenditures, inequalities, and exclusions of the past.

We will conclude with an appeal for more collaborative anthropology in every sense of the term. We need more anthropologists working together and working with other humans and nonhumans on the problems that matter most in this world. Those problems, like energy transition, are complex, massively scaled, and very often ill suited to critical and activist engagement by individual researchers. As scholars, we will better understand our present dilemmas and possible paths forward if we work together, whenever possible drawing on varying but complementary skills and forms of expertise in the pursuit of responses. As beings living on a damaged planet, what we already understand is that none of us can exit the Anthropocene on our own. The hyperindividualism of the past three decades, the capitalist empire building of the past two hundred years, the Northern privilege of the past five centuries, the monotheistic patriarchy of the past two thousand years, the agrilogistics of the past ten millennia—all of this, everything, will have to be remade if a global humanity is going to be reborn that will not be actively, constantly destroying its lifeworld and the lifeworld of the majority of the earth's species. This project will be utopian in the sense that it will have to make a world that has not yet existed. It will be revolutionary in the sense that it will not be accomplished by technology, or markets, or violence, or anthropocentrism, or any of the other behaviors and attitudes that brought us here in the first place. It will be a project accomplished by humans who can accept their own diminishment of importance and entitlement relative to their nonhuman neighbors and by those who are willing to work collaboratively to restabilize the vital systems of geos and bios on this planet. These are the politics, aeolian and otherwise, to which we should commit ourselves, these are the futures worth having.

Notes

JOINT PREFACE

1. See Lynch 1982; Price 2016.

2. For more information on these partnerships, see Ethnographic Terminalia, http://ethnographicterminalia.org; "Anthropology of the World Trade Organization," Institut interdisciplinaire d'anthropologie du contemporain, February 12, 2008, http://www.iiac.cnrs.fr/article1249.html.

3. But here, as in other respects, we find the aforementioned collaborative partnerships trailblazing. See, for example, Matsutake Worlds Research Group 2009; the exhibition catalogs and zines produced by Ethnographic Terminalia, http://ethnographicterminalia.org/about/publications; Abélès 2011.

4. See, for example, Boyer and Marcus, forthcoming.

INTRODUCTION

1. Latour 2004.

2. The counterfactual that is usually offered to offset the failure of the COP process is the success of the Montreal Protocol of 1989. However, it is worth mentioning that this protocol also inadvertently accelerated global warming by shifting from the industrial use of chlorofluorocarbons to hydrofluorocarbons, a process that has taken a further three decades to address.

3. On the crisis and/or compromise of neo/liberal political institutions see Brown 2015; Mouffe 2005; Rancière 1998, 2001; Sloterdijk 1988; Žižek 1999, 2002. Swyngedouw (2009) offers a perceptive analysis of the "postpolitical," technocratic character of environmental politics generally. Political anthropologists have recently begun to explore ironic responses to overformalized and performative modes of political practice, e.g., Bernal 2013, Boyer 2013b; Boyer and Yurchak 2010; Haugerud 2013; Klumbyte 2011; Molé 2013. These processes have meanwhile become a key focus of an

emerging anthropology of climate, e.g., Crate 2011; Crate and Nuttall 2009; Dove 2014; Fiske et al. 2014.

4. Povinelli 2016, 28.

5. Among many Marxist critics of neoliberalism, see especially Duménil and Lévy 2011; Harvey 2005, 2007; LiPuma and Lee 2004. For more Foucauldian takes on neoliberalism, see, e.g., Foucault 2004; Ong and Collier 2005.

6. On "integralism," see Holmes 2000.

7. Muehlebach 2016.

8. Boyer 2013b.

9. See, e.g., Transition Culture, last updated February 2017, http://transitionculture .org; Degrowth, https://degrowth.org; Kallis 2011, 2018.

10. Klein 2015; Graeber 2010.

11. Colebrook 2017.

12. Boyer 2014.

13. Hymes 1972.

14. Haraway 2003; Kirksey and Helmreich 2010.

15. See Tsing et al. 2017.

16. Stengers 2005; Lévi-Strauss 1966.

17. Morton 2013; Boyer and Morton 2016.

18. For additional reflections on the ethics of anthropological theorization today that inform my position here, please see Boyer 2010; Boyer, Faubion, and Marcus 2015.

19. This is the key premise of the argument for the concept of "Capitalocene" over "Anthropocene."

20. See Colebrook 2017.

21. Or, as Dipesh Chakrabarty has recently stated, "Today you need to both zoom out and zoom in. Unless you zoom in, into the finer resolution of the story, you don't see what humans are doing to each other. But if you don't zoom out, you don't see the human story as a whole in the context of other species, in the context of history of life." Cultures of Energy podcast, episode 19, June 10, 2016, http://culturesofenergy .com/ep-19-dipesh-chakrabarty/.

22. Derrida 1976.

23. See Boyer and Howe 2015 for a fuller discussion of the capacities and mobilities of anthropological knowledge.

24. Elizabeth Povinelli has made a similar argument concerning her neologism, "geontopower" (2016), arguing that it is not an effort to posit "a new metaphysics of power" but rather to "help make visible the figural tactics of late liberalism as a long-standing *biontological orientation and distribution* of power crumbles, losing its efficacy as a self-evident backdrop to reason" (italics original). A concept like "energopower" plumbs a similar rift as Northern biopolitical imagination becomes disrupted and disabled by ecological and geological forces unleashed by its very efforts to enhance life and civilization.

25. I will sidestep here the lively debate over the ontological status of objects and "correlationalism" in contemporary continental philosophy (Bennett 2009; Bogost 2012; Harman 2002; Meillasoux 2008; Morton 2013; Povinelli 2016). Although the

epistemic attentions of this discussion are lively, its political attentions seem more impoverished, which makes it, on the whole, less urgent for this discussion of enablement.

26. The crucial texts that outline Marx's philosophy of labor and capital are the 1844 *Manuscripts* (particularly the manuscript on "Estranged Labor"), *The German Ideology*, and the *Grundrisse*. See Marx (1861) 1974 and also https://www.marxists.org/archive/marx/works/1845/german-ideology/index.htm and https://www.marxists.org/archive/marx/works/1844/manuscripts/preface.htm.

27. Sutherland 2008.

28. Burkett and Foster 2006, 127.

29. See Marx (1861) 1974, notebook 7.

30. See Marx (1861) 1974.

31. Land 2011; Mackay 2014. See also Alex Williams and Nick Srnicek, "#Accelerate Manifesto for and Accelerationist Politics," May 14, 2013, Critical Legal Thinking, http://criticallegalthinking.com/2013/05/14/accelerate-manifesto-for-an-accelerationist-politics/.

32. Sunder Rajan 2012, 14.

33. Malm 2013; Malm and Hornborg 2014.

34. Foucault 1984, 143.

35. Among them, Agamben 1998; Deleuze 1995; Hardt and Negri 2000; Povinelli 2016; Rabinow and Rose 2006.

36. Among them, notably, Beck 2007; Biehl 2007; Briggs 2005; Cohen 2005; Ferguson and Gupta 2002; Franklin and Roberts 2006; Fullwiley 2006; Greenhalgh and Winckler 2005; Lakoff and Collier 2008; Petryna 2002; Redfield 2005; Sunder Rajan 2012.

37. Foucault 2000, 216–17.

38. See Foucault 1980.

39. Foucault 1979.

40. Foucault 1984, 143.

41. Rabinow and Rose 2006, 193.

42. See, e.g., Briggs and Nichter 2009; Greenhalgh and Winckler 2005; Petryna 2002.

43. Luke 1999; relatedly, Malette 2009.

44. Foucault 1993, 202.

45. Boyer 2014.

46. In anthropology, this thinking was epitomized by the work of Leslie White (1943, 1949, 1959). See also, as examples of broader public commentary on the promises and perils of atomic energy, O'Neill 1940; Potter 1940.

47. See, e.g., Adams 1975, 1978; Rappaport 1975, in the Whitean tradition alongside a robust literature on the cultural and social impacts of energy development for indigenous peoples (Jorgensen et al. 1978; Jorgensen 1984; Nordstrom et al. 1977), especially in terms of nuclear power (Robbins 1980), uranium mining (Robbins 1984), and oil extraction (Kruse, Kleinfeld, and Travis 1982; Jorgensen 1990). Laura Nader's research on the "culture of energy experts" (1980, 1981, 2004) was pathbreaking and

helped set the stage for more recent ethnographies of energy experts (Mason and Stoilkova 2012) as well as for political anthropologies of carbon (Coronil 1997) and nuclear (Gusterson 1996; Masco 2006) statecraft.

48. See, e.g., Foucault 1984; Haraway 1985; Latour 1988.

49. This not only attracted renewed theoretical attention but also generated a much larger wave of ethnographic interest among anthropologists who have helped to renew energy as a research thematic. See, e.g., Appel, Mason, and Watts 2015; Behrends, Reyna, and Schlee 2011; Crate and Nuttall 2009; Henning 2005; Johnston, Dawson, and Madsen 2010; Love 2008; Love and Garwood 2011; Mason 2007; McNeish and Logan 2012; Nader 2010; Powell and Long 2010; Reyna and Behrends 2008; Rogers 2015; Sawyer 2004, 2007; Sawyer and Gomez 2012; Smith and Frehner 2010; Strauss and Orlove 2003; Strauss, Love, and Rupp 2013; Wilhite 2005; Winther 2008.

50. Scheer 2002, 2006.

51. Scheer 2002, 89.

52. Boyer 2016.

53. Mitchell 2009, 2011.

54. Mitchell 2009, 407.

55. Mitchell 2011, 173–99.

56. See, e.g., Harvey 2007; Duménil and Lévy 2011.

57. This is one of Mitchell's central arguments: "When the global financial order was reconstructed after the Second World War, it was based not on reserves of gold, but on flows of oil. Gold reserves could no longer provide the mechanism to secure international financial exchange, because the European allies had been forced to send all their gold bullion to America to pay for imports of coal, oil and other wartime supplies. By the end of the war the United States had accumulated 80 per cent of the world's gold reserves. The Bretton Woods Agreements of 1944 fixed the value of the US dollar on the basis of this gold, at $35 an ounce. Every other country pegged the value of its currency to the dollar and thus indirectly to the American gold monopoly. In practice, however, what sustained the value of the dollar was its convertibility not to gold but to oil. In both value and volume, oil was the largest commodity in world trade. In 1945 the United States produced two-thirds of the world's oil. As production in the Middle East was developed, and the routes of pipelines plotted, most of this overseas oil was also under the control of American companies" (2009, 414).

58. See, e.g., Graeber 2011.

59. What I mean by "neoliberal disarticulation" is that petropower is no longer made to directly serve state interests as it was in the heyday of Keynesianism. Instead, the typical arrangement is an alliance between oil and gas corporations and certain political factions and institutions in order to allow petropower to be exerted on behalf of those corporations, their shareholders, and the speculative interests of the market.

60. See, e.g., Howe et al. 2015; Gordillo 2014; Gupta 2013.

61. Boyer 2016.

62. For Aristotle, "enérgeia" meant "activity" or "action" as distinct from δύναμις (*dynamis*), which meant "power" in the sense of capacity. James Faubion (pers.

comm.) cautions against "any temptation to read back the Newtonian notion of mechanical energy into Aristotle. . . . What happens is that energeia is taken up in Latin as meaning (basically) the force or vigor of expression (of words but also of potentiae [potentials], which allows it to be brought into the same semantic field with vis (force, power) and vis viva (living force, living power) and which basically allows the classical distinction between energeia and dynamis to collapse into the singular notion of energy/force as it is codified in mechanistic physics." Cara Daggett (2019) glosses Aristotle's enérgeia as "dynamic virtue" and strongly differentiates it from the Victorian conceptualization of energy as work, which was shaped by, among other forces, empire, evolutionary theory, Presbyterianism and thermodynamics.

63. Massumi 2015. Along similar lines, the work of Jane Bennett on "vibrant matter" comes to mind (2009), as does Povinelli's critique (2016) of the biontological premise of (Deleuzian) affect theory and vitalist thinking more generally.

64. Boyer 2013b, 152–56.

65. Žižek 1997.

66. As well as memories of the futures we wish to avoid; see, e.g., Oreskes and Conway 2013.

67. Many projects deserve recognition here. Those that have influenced this project most directly include the multimedia works of Brian Eno, Natalie Jeremijenko, Jae Rhim Lee, Smudge Studio, and Marina Zurkow as well as the "climate fiction" of Margaret Atwood, Paolo Bacigalupi, J. G. Ballard, Ian McEwan, Kim Stanley Robinson, Jeff VanderMeer, and Claire Vaye Watkins.

68. Howe and Boyer 2015.

69. I mean "terroir" here less in the specific sense of "soil" and more in the capacious sense of local "climate" and "environs." It refers to the mesh of local power forms and forces that give a situation its distinct character, which we are only able to fully understand by being in that context.

70. I thank one of the two anonymous reviewers of the duograph for encouraging me to highlight the importance of the politics surrounding communal land tenure in this volume of the duograph. As they wrote in their notes, "One of the major causes of internal community conflict is that a relatively small number of comuneros are the legal owners of the comunidad/ejido land (most but not all are men). The comisariado assembly is their collective space for decision making. The whole comunidad agraria apparatus is heavily linked to the Mexican state imagination of rural agricultural productivity. This is a kind of terroir. . . . A[nother] major axis of intracommunal tension is between the citizens who are not commoners/land owners and get no benefit from individual private sales/leases or even community land leases. The collective political space for these citizens is the municipality/agencia municipal. The municipio has very little tax money and may be trying to raise money for roads, schools, etc., through its limited land use change authorization policy." These forms of tension are discussed at greater length in chapter 1 in the matter of the bienes comunales of Ixtepec and in chapter 2 in the politics of the La Ventosan ejido and smallholder private landowners.

1. See Meyer (2004) and Christensen (2013) for discussions of the history of Danish wind power and the particular importance of community-owned wind installations in Denmark. The Yansa Group involved two entities, the Yansa Foundation and Yansa Community Interest Company (CIC). See the Yansa Group website, http://www.yansa.org/sample-page/.

2. The Asociación Mexicana de Energía Eólica (www.amdee.org) was founded in 2004. The president at the time of Sergio's visit was Carlos Gottfried Joy (see chapter 2).

3. Government records from the Registro Agrario Nacional indicate that the Ixtepecan bienes comunales were officially decreed on May 17, 1944, with control over 29,440 hectares (approximately 114 square miles) of land (http://www.gob.mx/ran). As in other Mexican communal estates, all decisions regarding land use are decided through the collective discussion and voting of the asamblea. The lands are administered by a three-person (president, secretary, treasurer) commissariat (*comisariado*), which is in turn supervised by a three-person *consejo de vigilancia* (oversight committee). Elections for both groups occur every three years.

4. When referring to the political entity governing the Ixtepecan bienes communales, Sergio tended to use the term "comuna" whereas the comuneros themselves typically referred to it as the "comunidad" or the "asamblea." I vary my own descriptions accordingly.

5. The worker-peasant-student coalition of the Isthmus of Tehuantepec, COCEI (la Coalición Obrera Campesina Estudiantil del Istmo), was founded in Juchitán in 1973. According to Howard Campbell, they "took power in Juchitán in 1981, after years of struggle in which more than 20 *Coceistas* died at the hands of *pistoleros*, police, and soldiers. Juchitán became the first Mexican city controlled by the Left since the emergence of the PRI and its immediate predecessors (ca. 1929). After two turbulent years of COCEI rule, during which time bloody shootouts and street fights were commonplace, the Mexican government invaded City Hall, imprisoned numerous *Coceistas*, established a state of siege in Juchitán, and restored PRI to power. Although COCEI was temporarily forced to go underground, Juchitán became a cause celebre among urban leftists and intellectuals, which sheltered the movement to some degree from further repression" (Campbell 1990, 50). See also Campbell et al. 1993; chapter 5 of this volume.

6. Anthropology has had its own long and complicated relationship with development (e.g., Foster 1969; Malinowski 1929; Mead 1953) that deserves mention since it is an important context for this ethnographic intervention. The expansion of programs aimed at fostering international aid and local development after the Second World War helped establish a new subfield of "applied anthropology" (Evans-Pritchard 1946) that rapidly became an important area of professional anthropological engagement during the 1950s and 1960s (Edelman and Haugerud 2005, 6–7). Concerns about the political conservatism of applied anthropology multiplied in the late 1960s. New paradigms such as "dependency theory" and "world systems theory" viewed the "underdevelopment" of the "third world" as a product of Western capitalism and imperialism's efforts to maintain structural forms of dependency and domination in

its postcolonial peripheries (Frank 1969; Wallerstein 1974, 1979). Critical paradigms of transnational relations had a strong impact in anthropology in the 1970s and 1980s (see, e.g., Nash 1981; Roseberry 1988; Wolf 1982), and Kearney (1986, 338) notes especially the impact of dependency theory in anthropological research on development in Latin America (e.g., Kemper 1977; Lomnitz 1977; Wiest 1978). An important consequence of this critical turn in cultural anthropology was that it created the opportunity for greater anthropological investigation of the institutions, practices, and forms of expert knowledge involved in national and international development programs. Scholars such as Escobar (1991, 1994) and Ferguson (1990) took up this opportunity in the early 1990s. Although theoretically opposed in many respects to critical political economy approaches, the political anthropology of development shared a skepticism for the modernist ambitions of development programs (cf. Apffel-Marglin and Marglin 1990; Mueller 1986) and highlighted how apparently depoliticized technical projects of intervention could actually "end up performing extremely sensitive political operations involving the entrenchment and expansion of institutional state power almost invisibly" (Ferguson 1990, 256). The vibrant anthropological research literature focused on development in the neoliberal era (e.g., Bornstein 2005; Hanson 2007; Li 2007; Rudnyckyj 2010; Sharma 2006; Vannier 2010) has developed this Foucauldian model further, paying special attention to the contributions of corporations and nonstate political entities such as NGOs, religious organizations, and social movements to "neoliberal governmentality."

7. Fisher (1997) offers a helpful introduction to early anthropological writing on NGOs, highlighting that "the growth of a multicentric world and the practices of growing numbers of nonstate national and transnational actors have had significant impact on the sites and communities that have been the focus of anthropological research" (459). Anthropologists have subsequently been particularly attentive to the "statelike" functions that NGOs absorbed during the heyday of structural adjustment policies during the 1990s (see, e.g., Elyachar 2005; Karim 2001; Smith-Nonini 1998) as well as to the complexity of NGOs' mediating relationship to social change in the neoliberal era (e.g., Grewal and Bernal 2014; Richard 2009; Schuller 2009; Redfield 2005). Yansa, in our experience, was quite sincere in its respect for indigenous binnizá traditions of land use and management. Yet it also wished to modernize these traditions in such a way as to undermine traditional patriarchal hierarchy and privilege, thus better commensurating the terralogical order of the comuna with modern liberal notions of belonging and communicative rationality (e.g., Habermas 1989). As fellow liberal subjects, we often felt that Yansa's intervention had the potential to bear positive fruit. But it was clear enough that their cultivation of "community" was a power-knowledge anchored to a particular understanding of social development that was by no means "indigenous" to southern Mexico.

8. In the sense that MacKenzie has argued, following J. L. Austin, that expert models can help to constitute a reality that they purport to simply depict (see MacKenzie 2008; MacKenzie, Muniesa, and Siu 2007). Bourdieu's concept of "theory effect" is relevant here (Bourdieu 1991) as well as Barad's rumination on the need to expand performativity beyond the scope of the human (Barad 2003).

9. Roughly speaking, Yansa's proposal was to organize funding for materials while relying on the men of El Zapote to provide the heavy labor of digging and laying the pipework. This led to some question about how men who had to work their fields could find the time to hold up their end of the bargain.

10. Grupo Yansa 2012, 15.

11. Grupo Yansa 2012, 10. See also Hoffmann 2012.

12. The question of land ownership in the isthmus is a highly complex and contentious issue that will receive fuller ethnographic treatment in chapters 2, 4, and 5. The short version is that Article 27 of the Mexican Constitution of 1917—what Ferry has termed "the most symbolically important piece of Mexican legislation of the past 140 years" (2005, 206)—guarantees unimpeded usufructuary land rights and considerable local sovereignty to the asambleas governing bienes comunales and bienes ejidales and even the right to recover traditionally owned land away from large colonial estates. But Article 27 also states that all land is ultimately part of national patrimony and subject to national law, which means leaving open the possibility of revoking usufructuary rights and asamblea sovereignty "as the public interest may demand" ("Mexico's Constitution of 1917 with Amendments through 2007," Constitute, https://www.constituteproject.org/constitution/Mexico_2007.pdf). This keeping-while-giving, part of the mestizo "cunning of recognition" (Povinelli 2002), has clouded questions of land tenure in Mexico ever since. Between the 1930s and the 1970s the Mexican federal government transferred fully half the country's land base to communal management (Perramond 2008). In Oaxaca, the impact was still higher, with 823 ejidos holding 18 percent of the total land in the state and 716 comunidades holding 67 percent of the state's land (Brown 2004, 4), empowering smallholders in areas where indigenous agrarian nuclei were especially widespread and politically active. In 1992 counterreforms prohibited further expansion of communal lands and opened bienes ejidales to the possibility of privatization. Although some scholars and many critics interpreted the 1992–93 reforms as part of a wholesale neoliberalization of Mexican society (e.g., Gledhill 1995a; Hellman 1997), others such as Perramond (2008) have argued that the land reforms were less impactful than is often assumed and that, in the end, "the ejido has not been subsumed under the new fabric of market-led approaches" (see Haenn 2006 for another perspective). In any event, slippage between usufructuary rights and formal rights to land, clouded further by the federal government's ultimate right of expropriation, is a crucial dynamic in all that follows.

13. Distrito Federal, or DF, the former name for the federal district of Mexico City, was dropped in 2016.

14. The building includes both the auditorium in which the general assembly of the bienes comunales meets as well as the offices of the comisariado.

15. In the spirit of Fabian 1983.

16. On the "particularly vexed" relationship between electric power and the poor in the global South, see Gupta 2015. Gupta's comments on the "precarious power" of illegal tapping of energy infrastructure (2015, 561) are also very apt in the context of southern Mexico, where the resources of both the grid and pipelines are subject to various practices of informal "recovery." Chance (2015) likewise sheds light on the intersection of energy, protest politics, and poverty in the South.

17. The enabling relationship of electricity to development and modernity is becoming a lively area of anthropological inquiry (see, e.g., Anusas and Ingold 2015; Boyer 2015; Degani 2017; El Khachab 2016; Gupta 2015; Özden-Schilling 2015), building upon the work of anthropologists such as Tanja Winther (2008) and Hal Wilhite (2005) and historians such as Thomas Hughes (1983) and David Nye (1992, 2010).

18. There is widespread anger about the cost of electricity in southern Mexico, which has spawned several civil resistance movements against *las altas tarifas* (the high tariffs). During the course of fieldwork, we heard many accusations that CFE was possibly illegally and certainly immorally raising tariffs on the isthmus as a new mode of exploitation. The company denies such claims fervently (see chapter 4 for their explanation of why istmeño electricity bills are rising), but at the very least, they could be accused of having done a very poor job of explaining how electricity is priced and how certain kinds of appliances would affect billing.

19. See chapter 4 for a fuller discussion of CFE's electricity pricing model. Mexico actually has among the highest domestic electricity subsidies anywhere in the world. Industry experts frequently describe the generosity of its subsidy structure as the reason that CFE is essentially bankrupt and unable to invest significantly in the power development projects it needs to meet consumer and industrial demand.

20. See Lomnitz 2005 on negative reciprocity.

21. Daniel is referring to the PROCEDE program of land registration (see de Ita 2003; Tiedje 2008).

22. See Kelly et al. 2010. The legal distinction therein is that the federal government granted stewardship of land to ejidos following the Mexican Revolution whereas with comunidades it recognized the stewardship they already possessed. See also Cornelius and Myhre 1998; Haenn 2006; Castellanos 2010.

23. The specific articles to which Daniel referred were "los artículos 1, 2, 14, 16 y 27, fracción VII de la Constitución Política de los Estados Unidos Mexicanos; artículo 21 de la *Convención Americana sobre Derechos Humanos*; Artículos 23 fracción X; 26, 93, 94, 95, 98, 99 y 100 de la *Ley Agraria*; así como lo dispuesto por los artículos 6, 13, 14, 15 y 16 del *Convenio 169 de la* OIT *relativa a Pueblos Indígenas y Tribales en países independientes y las disposiciones de la Declaración de las Naciones Unidas sobre Derechos de los Pueblos Indígenas*."

24. This majority was difficult to manage, in part because hundreds of deceased comuneros remained officially on the rolls.

25. On the dynamics of mestizaje nationalism in Mexico—which sacralizes the intermixing of indigenous and European traditions for creating a nation superseding its ancestries—past and present, see, in particular, Alonso 2004; Hartigan 2013; Loewe 2011; Lomnitz 2001; Stephen 1997; Stern 2002.

26. Published in the *Gaceta Parlamentaria*, no. 3627-3, October 18, 2012. See also Méndez and Garduño 2012.

27. Southern Mexico—or what in the logic of the CFE grid is termed the "Área Oriental"—is composed of eight states: Guerrero, Morelos, Puebla, Tlaxcala, Veracruz, Tabasco, Oaxaca, and Chiapas. The Área Oriental was a net electricity producer for Mexico even before the introduction of the wind farms largely because of the Grijalva hydroelectric complex in Chiapas. According to CFE planning documents, in 2011 the

peak demand for these eight states was 6,577 megawatts. At the same time, the area's generation capacity was nearly twice that at 12,856 megawatts.

28. According to a spokesperson of its manufacturer (ABB), the SVC device at Ixtepec Potencia "provide[s] fast-acting reactive power compensation in high-voltage electricity networks. This enhances stability by countering fluctuations in the voltage and current of an electric grid, thereby allowing more power to flow through and at the same time maintaining safety margins and increasing network stability."

29. The first temporada abierta ran from 2010 to 2014 and brought nine wind park projects totaling 1.891 gigawatts of new generation capacity as well as 145 kilometers of new transmission lines and, of course, the Ixtepec Potencia substation on grid.

30. As Norget writes, *malinchismo* refers "to a willingness to sell out your own, an attitude of disdain toward your own country, people or goods" and is often racialized, involving the "denigration of a dark-skinned indigenous identity," which is taken as "an affront to mestizo Mexicans" (2006, 57).

31. The other amparo was resolved by default. Ixtepec was initially able to win an appeal of the judge's initial decision to reject their Sureste I Fase II amparo. But ENEL Green Power was eventually allowed to complete the building of its park. Sureste I Fase II was inaugurated in March 2016 by President Enrique Peña Nieto, who spoke of it as proof of the "great benefits" produced by his controversial energy reform law.

32. The Procuraduría Agragria is Mexico's rural property regulatory agency.

33. In July 2016, Sergio told us that several Maya communities in Yucatán had also expressed interest in pursuing the development of community-owned wind parks.

2. LA VENTOSA

1. For further discussion of the impact of the PPP regime across the world, see Kuriyan and Ray 2009; Shore and Wright 2003.

2. Approximately $325 per hectare. Project financiers we later interviewed in Mexico City spoke of a gentle upward trend in land rent payments since the wind boom began in the mid-2000s. Comparatively, compensation for land usage in the United States tends to focus on some combination of the number of turbines installed, the number of megawatts produced, and a percentage of net profit from electricity sales. The variability and privacy of contracts on all sides make it difficult to compare whether Oaxacan compensation for land use is on par with, for example, the United States. But it is indeed plausible, in the best-case scenario outlined by Fernando, that a Oaxacan landowner might receive roughly equivalent annual payments to those of an US landowner. Still, in Germany, the estimated average range is between 50,000 and 70,000 Euros per year per turbine, which is several times higher than the Oaxacan payments. See, e.g., Coerschulte 2014.

3. Diódoro Carrasco Altamirano (1992–98), José Murat Casab (1998–2004), and Ulises Ruiz Ortiz (2004–10).

4. On the impact of Mexican land reform on bienes ejidales, see Perramond 2008. See also discussion by D. Smith et al. (2009) of the land registration program

PROCEDE, especially regarding Oaxaca's low participation rate and tendency toward strategic partial engagement of land reform.

5. These "smallholders" historically suffered greatly from land expropriation during the Porfiriato (the years when Porfirio Díaz and his allies ruled, 1876–1911; see Katz 1988) and were among those to be reenfranchised by the land reforms promised by the Mexican Constitution of 1917. Joseph (1988, 122ff.) argues that the revolutionary idea of pequeña propiedad, as enshrined in the Constitution, captured the debates over bourgeois land reform that sought to replace the haciendas with modernized land relations (and modernized indigenous communities). However, as Hu-DeHart (1988, 175) observes in the Sonoran context, by the latter half of the twentieth century, the term had often come to include what were, in fact, very large and powerful landowners. Porfirio Montero in La Ventosa is a pequeño propietario in this latter sense.

6. Most often translated as "bosses," these informal local powerbrokers are seemingly as ubiquitous as they are various across the Mexican political landscape. Pansters explains that their essential political character is hard to pin down analytically, belonging to a "a semantic cluster that incorporates a number of other notions such as patronage, (inter)mediation, hierarchy, informality, violence, territory, authoritarianism, but also leadership, consent, paternalism and corruption" (2005, 350). Caciquismo involves coercive power, which is sometimes violent, but also the normative power of brokerage and efficacy (Friedrich 1965, 1970). See Bartra and Huerta 1978; Greenberg 1997; Guerra 1992; Knight and Pansters 2005; B. Smith 2009, for more detailed discussion of the roots and consequences of caciquismo in Mexico. The term "cacique" also carries with it a Taíno indigenous etymology, meaning that it also signals a premodern racialized form of authority that is viewed as incommensurable with the principles and institutions of legitimate settler-mestizo governance.

7. The COCEI uprising rekindled and transformed an older political division between "red" and "green" factions, which in the early part of the twentieth century corresponded to class tensions in Juchitán between the northern (more elite, "red") and southern (more working class, "green") wards of the town. When the federalists burned Juchitán in 1911 to quell the (green) Chegomista rebellion, this was viewed as a victory for the reds, driving many greens out of Juchitán to nearby villages including La Ventosa. There green/red anger and resentment simmered for decades. In the 1980s, with the rise of COCEI to power in Juchitán, Montero and his allies cleverly invoked COCEI's political redness to ignite local fears that the Juchiteco elites were again coming to trample La Ventosan interests—even though those these reds no longer identified with conservative elite social forces, indeed quite the contrary. Montero's green victory over the COCEI made him something of a local legend, lending him charismatic authority in La Ventosa for years afterward that helped secure his cacicazgo. Ristow (2008) and Rubin (1997) relate the major elements of this story, although Montero's own story was one we had to piece together through oral accounts and the help of José López de la Cruz.

8. It was often said that Montero had an uncanny knack for knowing where wind parks were going to be situated and for buying out other private landholders in advance.

9. An Iberdrola representative estimated in December 2011 that the company was paying roughly seven million pesos a year for eighty megawatts. Given those numbers, Montero's proposal would have represented a 12.5 percent increase in payments.

10. An investigation of the social impacts of wind park development in the isthmus undertaken by the Centro de Investigaciones y Estudios Superiores en Antropología Social (CIESAS) and led by Salomón Nahmad Sittón found that in La Ventosa the issue of variable payments had indeed created significant tensions. Specifically, Iberdrola's initial plan to pay 1 percent to 1.3 percent of the revenues from generation made "those with contracts with Iberdrola complain that the payment for generation is less than fixed payments; they feel deceived. . . . They always have the idea that they are being robbed. . . . [And even with the promise of new contracts from Iberdrola,] the lack of information has already generated much distrust" (Nahmad Sittón 2011, 66).

11. For insight into the complex history of energy and power before, during, and after the 1970s oil shocks, see, e.g., Bini and Garavini 2016; Dietrich 2008; Mitchell 2011; Venn 2002.

12. Breglia (2013, 5) writes, "Mexican oil is a foundation for the economy and a cornerstone of nationalism. Lifeblood for the national economy, crude oil is strategically important for the domestic economy. Its revenues fund up to 40 percent of the national budget. In addition to supporting the energy needs of its neighbor the United States through exports, oil also serves Mexico's own ever-rising domestic energy demand. These twin pressures on Mexico's most valuable asset place the issue of resource sovereignty . . . at the forefront of contentious debates."

13. See UNIDO 1975, p. 2, par. 10.

14. UNIDO 1975, p. 11, par. 58, secs. f, g, l.

15. IIE was part of the Mexican Ministry of Energy, SENER. See Caldera Muñoz et al. 1980. It was succeeded by the Instituto Nacional de Electricidad y Energías Limpias (INEEL).

16. Aiello et al. 1983.

17. Borja Díaz et al. 2005, 41.

18. Lomnitz writes of the Mexican state's "dismodernity" during this period: "The 1982 debt crisis dealt a terrible blow to the regime of state-fostered national development, and the economic arrangement that has emerged provoked an intense struggle for supremacy between diverse modernizing formulas" (2001, 111).

19. In terms of electrical infrastructure specifically, CFE's power generation capacity more than doubled from 7.41 gigawatts in 1970 to 16.85 gigawatts in 1980. But in the subsequent decade, capacity only increased very gradually to 18.27 gigawatts (Breceda 2000, 11).

20. The consequences of this "turn" deserve some further discussion. Although anthropologists (among others) have theorized the general political, social, and cultural impacts of the global rise of neoliberal politics and policy (e.g., Comaroff and Comaroff 2001; Ferguson and Gupta 2002; Gledhill 2007; Harvey 2005), political anthropology of Latin America has been especially successful at demonstrating that economic globalization and neoliberalism must be viewed as heterogeneous phenomena with complex and at times seemingly paradoxical cultural impacts. The social and

cultural conditions of the neoliberal era in Latin America have been shown to combine, for example, new emphases on militarization and rule of law with rising levels of everyday violence and insecurity (e.g., Goldstein 2004, 2005; Rochlin 1997; Wacquant 2008), overtures to participatory democracy with resurgent charismatic and populist politics (e.g., Paley 2001; Schiller 2011), and intensified ethnic and racist differentiation with political multiculturalism (e.g., Hale 2006; Postero 2006). In the case of Mexican neoliberal statecraft and political culture, Schwegler has argued that "we lack a nuanced understanding of the mechanisms, processes, and practices by which neoliberalism becomes entrenched in the institutions credited with its dissemination" (2008, 682). This is a striking omission, especially given anthropologists' comparative attention to other important dimensions of neoliberalism's influence upon contemporary social experience in Mexico, including consumption and middle-class identity (Cahn 2008), national mythology and imagination (Bartra 2002; Lomnitz 2001), labor and migration (Cohen 2004; Kelly 2008; Wilson 2010), rural poverty (Gledhill 1995b), land reform (McDonald 1999; Nugent and Alonso 1994; Castellanos 2010), and the rising influence of NGOs as intermediaries between the Mexican government and communities (Richard 2009). Throughout this literature, the Mexican state is frequently portrayed as a key agent of neoliberal transformation and of the "growing deterioration of living standards" (Ochoa 2001, 154) within the country, whether through benign neglect or, it is more often argued, through active dismantling of social services, endemic corruption, and militarization.

21. Consejo Nacional de Ciencia y Tecnología, Mexico's national scientific and technology foundation.

22. These speeds equal 20.8 and 67.1 miles per hour respectively. The Isthmus of Tehuantepec is widely regarded in the industry as having some of the best resources for terrestrial wind power anywhere in the world.

23. See Schwegler 2008 for a nuanced account of these dynamics.

24. Declining petroleum production since 2004 created a parallel sense of political urgency around stimulating private investment in the oil and gas sector. In December 2012, Enrique Peña Nieto's resurgent PRI party was able to organize an alliance, the Pacto por México, with the major opposition, the PAN and PRD parties, in order to streamline reforms in the education, telecommunications, banking, and energy sectors. Among other reforms, Articles 25 and 27 of the Mexican Constitution were amended to mandate the transition of Pemex and CFE to "state-owned productive enterprises" charged with creating economic value and profits rather than simply serving as stewards of national resources. The PRD dropped out of the Pacto in 2013 in part because of the issue of private sector investment in oil; the reforms stressed public-private partnerships in oil and gas exploration going forward as well as opening mid- and downstream oil and gas to private investment. Nevertheless, the reforms passed into law in 2014, marking what is certainly the most radical shift in Mexican energy policy since the 1930s. The reforms have been celebrated outside of Mexico as long-overdue market liberalization guaranteed by "courageous leadership and governance" (see Goldwyn, Brown, and Cayten 2014, 37).

25. See, e.g., Breceda 2000.

26. See Breceda 2000, 24–27.

27. See Comisión Reguladora de Energia, https://www.gob.mx/cre. Critics of the CRE, such as the Frente de Trabajadores de la Energía, assert that the privatization of electricity generation is being hidden and that current and planned projects will raise the proportion of privately generated electricity to more than 66 percent of the national supply. See "Peña Nieto quiebra a Pemex y CFE," *Energía* 16, no. 335, May 15, 2016, http://www.fte-energia.org/E335/04.html.

28. Galbraith and Price 2013, 66–68.

29. "Enredos burocráticos congelan central en Zacatecas," *La Jornada*, May 1, 1998, http://www.jornada.unam.mx/1998/01/05/zacatecas.html.

30. The permits were for Fuerza Eólica del Istmo, completed in two stages of fifty megawatts and thirty megawatts in 2011 and 2012 respectively and for the La Rumorosa ten-megawatt park in Baja California, which was completed in 2010.

31. Capacity factor is the average power generated by a power installation as a percentage of nominal total capacity. The La Venta figures remain outstanding twenty years later. NREL's Transparent Cost Database lists the median plant capacity for onshore wind farms from 2010 to 2014 as 38 percent (with a maximum of 52 percent). See "Transparent Cost Database," OpenEI, http://en.openei.org/apps/TCDB/#blank.

32. He was then serving as director of micro, small, and midlevel enterprises for the Oaxacan state agency of industrial and commercial development (SEDIC),

33. There are also widespread rumors that Fernando's Fundación accepted donations from wind developers in exchange for helping them to achieve results on the ground in the isthmus.

34. Mimiaga Sosa 2009.

35. SEGEGO is the Secretaría General de Gobierno, the Office of the Secretary General.

36. Oceransky notes, for example, that a former Fundación employee, Alvarez Velasquez Maldonado, went to work for the Spanish firm Preneal, which then received several choice plots in the core wind zone in the first temporada abierta (open season for investors) despite never subsequently developing a single project (2010, 513–14).

37. See Elliott et al. 2003.

38. See Fernández-Vega 2004.

39. See the *Ecologics* volume.

40. José López de la Cruz pers. comm., 2013.

41. López de la Cruz pers. comm.

42. Velas are a pre-Hispanic ceremonial tradition practiced throughout the isthmus that have evolved into a series of annual public celebrations sponsored by different local associations, "comprising three days of parades and parties, an all-night dance, and a celebratory mass" (Royce 1991, 51). Velas are not only important markers of the social calendar of istmeño towns but also powerful identifiers of associative solidarity and, indeed, "community" itself (see Münch Galindo 2006).

43. See Ristow 2008, 403.

44. Rubin, for example, describes Charis as the "regional political boss" (1997, 45) of the isthmus from the 1930s until his death in 1964, striking a deal with Mexico City for

"regional autonomy in return for support for the post-revolutionary state" (1997, 48). See chapter 5.

45. More specifically, Rubin argues that although General Charis served the PRI cause loyally in federal matters, he impeded the development of local PRI networks in the area since they challenged his personal authority (1997, 53–54). Thus, it was not until the 1960s and 1970s that political parties were first able to build strong independent networks in the isthmus.

46. Binford (1985) analyzes this complex and highly consequential story in great detail.

47. See Campbell et al. 1993 on the rise of COCEI.

48. In recent years, the @ symbol has been used for gender inclusivity in contemporary Mexican Spanish, e.g., "l@s Méxican@s."

49. The La Ventosa ejido was decreed in 1954 as containing three thousand hectares, of which half is commons and half is currently parcelada into individual plots. There are 250 ejidatarios currently.

50. These two parks faced blockades and conflict in mid-2013 when a group of propietarios sought to derail the project because of a conflict over payments.

51. The payment for this easement was quoted to us by Cain López Toledo as seven thousand pesos per hectare annually. But it was not clear how many of the propietarios received it.

52. López de la Cruz is referring to UCIZONI (https://www.facebook.com/ucizoni/) and their leader, Carlos Beas.

53. There are striking resonances of istmeño knowledge of wind turbines to the "occult economy" described by Comaroff and Comaroff (1999).

54. For "Domingo" and "Ximena."

55. See, e.g., *El Sur Diario*, http://www.elsurdiario.com.mx/index.php?option=com _content&view=article&id=18767:-inauguran-centro-cultural-bacusa-gui-en-la -ventosa&catid=46:region&Itemid=95.

3. OAXACA DE JUARÉZ

1. On historical and recent dynamics of the PRI and PRIismo in Mexico, see, e.g., Adler-Lomnitz, Salazar-Elena, and Adler 2010; Aitken et al. 1996; Varela Guinot 1993; Vázquez 2003.

2. Clientelism, corporatism, and vertical integration were crucial vectors within the PRI party-state. Selee writes, "The creation and consolidation of the PRI as the party of power, together with the centralization of government functions and resources, successfully replaced the centrifugal forces of regional power holders with a functional structure for negotiating differences among political actors. . . . Within the PRI, local organizations—whether municipal party committees, neighborhood organizations, or union locals—were linked upward into ever larger second- and third-tier organizations that ultimately were part of one of the party's sectors. Clientelism—the unequal exchange of political support for public benefits—linked citizens to particular organizations and to their leaders at the local level. Corporatism within the national

and state PRI connected these organizations within vertical party hierarchies" (2011, xx). Although perhaps less disciplined and bureaucratic in terms of party ideology, the general structuring of the political order is indeed quite similar to the party-states that flourished elsewhere in the mid-twentieth century under the influence of Marxism-Leninism (see, e.g., Boyer 2005; Verdery 1996). The folding of charismatic power and clientage into bureaucratic authority is, of course, familiar outside of Latin America too. Hull's description of "Zaffar Khan" in Karachi is an excellent example (Hull 2012, 77–80).

3. On the APPO uprising and its repression, see Howell 2009; Martínez Vásquez 2007, 2008; Nahmad 2007; Norget 2010; Stephen 2013.

4. A standby letter of credit is a guarantee issued by a bank to function as a payer of last resort should their client fail to fulfill a contractual commitment with another party.

5. There are actually two teachers' unions at work in Mexico, the SNTE (Sindicato Nacional de Trabajadores de la Educación), founded in 1949 and currently the largest teacher's union in the Americas, and the CNTE (Coordinadora Nacional de Trabajadores de la Educación), founded in 1979. Both unions have challenged neoliberal reform and federal power in Mexico since the 1980s. In recent years, both unions' section 22 (the Oaxacan sections) have tended to work together against efforts from Mexico City to promote education reform, which the unions view as an effort to weaken their considerable political influence. The Oaxacan sections are known nationally for their militancy, particularly in the aftermath of the APPO uprising of 2006. We found that many of the key figures in the antieólico movement against wind megaprojects in the isthmus had ties to the SNTE/CNTE. One teacher told us proudly that the center of gravity of political resistance in Oaxaca had shifted from Oaxaca City to the isthmus after the decline of the APPO. He told us that the wind parks had become a rallying point for not only local resistance but also regional and national resistance to neoliberal capitalism through the work of Section 22.

6. "Sexenio" refers to the six-year period of a Mexican presidency or governorship.

7. See the *Ecologics* volume.

8. Paso de la Reina is actually a federal hydroelectricity project sponsored by CFE. More than a decade in the planning, the project has met fierce and sustained local resistance, including a community blockade since 2009. For documentation of the resistance, see Consejo de Pueblos Unidos por la Defensa del Río Verde, http://pasodelareina.org.

9. See the discussion of the history of Mexican wind power in chapter 2. Abardia's comment reflects the post-1982 consensus that private-public partnerships (PPPs) are the optimal way to attract capital and share risk in major infrastructure projects.

10. SEGEGO is the Secretaría General de Gobierno (focusing on resolving social and political conflicts), and SAI is the Secretaría de Asuntos Indígenas (charged with managing indigenous affairs) for the state of Oaxaca.

11. Gross's work on the rise of Protestantism in Oaxaca is very relevant here (2012a, 2012b). Evangelical Protestants now make up more than 10 percent of Oaxaca's population, which is higher than the national average in Mexico. In the isthmus, because of Porifirio Montero's powerful network, Evangelical Christianity has been a significant

force in the moral definition and social organization of wind power development. The Evangelical prioritization of self-interest over community interest is evident in Montero's land-grab tactics as well as in Casillas's opposition to corporatist politics. But the thrust of Evangelical individualism is often blunted by the dominant social Catholicism of Oaxaca, which in the isthmus is further inflected by indigenous communalism and by the clientage-network politics of caciquismo.

12. What came across more strongly in other conversations we had with Casillas was his spiritual commitment to environmental defense and maintenance. This was not, perhaps, the ecocentric-spiritual ethics that Bron Taylor has diagnosed as "dark green religion" (2010) but certainly seemed to us a sincere commitment to "creation care."

13. Cabrero Mendoza 2005, 5; see also 2007. The struggle between centralized and decentralized governmental authority in Mexico dates back to the independence struggle of the early nineteenth century (Salinas Sandoval 2014). Mecham has summarized the first hundred years (roughly) of the history of Mexican federalism: "Federalism has never existed in fact in Mexico. It is an indisputable commonplace that the Mexican nation is now and always has been federal in theory only; actually, it has always been centralistic" (1938, 164). The heyday of mid-twentieth-century PRIismo only reinforced this trend through its party organization: centralized bureaucratic authority and decision-making power remained concentrated in Mexico City. In 1957, Tucker commented, "Centralization of tax power has been growing steadily through the years, mainly through specific reforms to the Constitution" (160). Merchant and Rich have argued more recently that "the pattern for decades has been the subservience of governors and of state legislatures to Mexico City and the calculated denial of resources for genuine federalism" (2003, 663). At the same time, Loaeza describes how, in the wake of the structural adjustment reforms implemented by the Salinas and Zedillo regimes, "the weakness of the Mexican state has become a feature of the prevailing institutional arrangement" (2006, 34; cf. Selee 2011). That she does not differentiate strongly between the federal and state levels of "the Mexican state" suggests that the neoliberal era has been accompanied by a dissolution of federal authority without any substantial improvement at the state level.

14. This is an observation that has become a commonplace in political anthropology (see, e.g., Crewe and Axelby 2013; Das and Poole 2004; Sharma and Gupta 2006; Trouillot 2001). Yet, as Abrams has argued, the concept of "the state" is precisely that which conceals "the actual disunity of political power" (1988, 79) and, as such, has an important performative effect on political knowledge.

15. Although unnamed, the diputad@s clearly had in mind UCIZONI (https://www .facebook.com/ucizoni/) and the Asamblea de los Pueblos Indígenas del Istmo de Tehuantepec en Defensa de la Tierra y el Territorio (https://tierrayterritorio.wordpress .com), which were the most active NGOs working in opposition to wind park development in the isthmus.

16. Ruiz is a widely maligned and unpopular politician outside of the PRI (see, e.g., Stephen 2013).

17. As elsewhere in Mexico, PRI political networks in the isthmus are anchored by personal and kin loyalty to key figures and families. PRIista clientelism is frequently

indistinguishable from caciquismo, not just in local reckonings but also in the scholarly literature (see, e.g., Foweraker 1990; Kaufman 1981; Knight and Pansters 2005).

18. See the *Ecologics* volume.

19. See, e.g., Povinelli 2011; Jusionyte 2015. Late liberalism's contemporary condition often invites both emic and etic rhetorics of tragedy and crisis, emphasizing often the perversion of authentic liberal ideas and institutions by the logic of markets and capital (e.g., Brown 2015; Chalfin 2010; Harvey 2005; Muir 2015). But the parlous state of liberalism and democracy has also invited ironic and even ludic appropriations, reactions, and squattings (Bernal 2013; Boyer and Yurchak 2010; Boyer 2013b; Haugerud 2013; Knight 2015; Klumbyte 2011; Molé 2013) that both draw attention to the overformalization of late-liberal political practice but also to the potential for creative, even sincere, reappropriations of late-liberal ideas, values, and institutions.

20. Bobick 2012, 2014.

21. In his study of the administration of Oaxacan forests, Mathews (2011) argues that such performances are a more general condition of Mexican state-making, and indeed are characteristic in part of all state-making. "Once we see knowledge and nonknowledge as going together, the importance of thinking about knowledge-making and state-making as more or less dramatic and unstable performances becomes clear. . . . In the Mexican case, this means that officials have to pay great attention to the texture of their performances of authoritative knowledge. They have to pay attention to how the audience knows, to the audience's expectations of how state authority is to be performed, how knowledge is to be declared, and what forms of opposition and public declarations the audience might make." See also Jusionyte 2015 on this point.

22. Cochineal was Oaxaca's main source of income between 1758 and 1817 (Covarrubias 1946, 213).

23. UTVCO is a TSU (Técnico Superior Universitario) founded in 2009 and located in the Oaxaca Valley town of San Pablo Huixtepec. It offers a course of study in renewable energy with a special focus on solar energy. DAAD is the German Academic Exchange Service.

24. See the *Ecologics* volume.

25. The wholesale market price of electricity in Germany declined more than 66 percent in five years (2011–16) owing largely to the oversupply of electricity created by rapid renewable energy development combined with a slowness to retire conventional baseload bulk power supply. See, e.g., Flauger and Hubik 2016.

26. A partly derogatory term for someone from Mexico City.

27. On UCIZONI, see Kraemer Bayer 2008, 137ff.; Zafra, Hernández-Díaz, and Garza Zepeda 2002.

28. Boyer 2003, 538.

29. The Guelaguetza festival has attracted significant anthropological attention as a site for exploring the complexities of indigeneity, race, and performance in Oaxaca. See, e.g., Brulotte 2009; Chavez 2013; Goertzen 2010; Kearney 2004; Lizama Quijano 2006; Royce 1991, among others.

30. Since 1995, Oaxacan municipios have been able to run elections through a communal assembly (normally comprising predominantly if not exclusively men) instead

of a ballot box (Clarke 2000, 168). Benton (2011) calculates that a strong majority (418 of 570) of Oaxacan municipios have adopted the usos y costumbres model of governance and argues that the popularity of the shift can be linked to a desire to maintain social stability with the deterioriation of the authority of the PRI party-state.

31. Sánchez Prado (2015) has written forcefully of the "impossibility of the political" in Mexico, arguing that "that the very question of the future of radical politics in Mexico must start with a substantial reimagining of a future that would disrupt the role of the rule of law as master signifier and the liberal matrix that supports it." On the limits of settler liberalism, see also Burke 2002; Povinelli 2002; Simpson 2014; Veracini 2010.

32. Ahmed 2014; see also Ahmed's FeministKilljoys blog post, "Against Students," June 25, 2015, https://feministkilljoys.com/2015/06/25/against-students/.

33. Stevenson 2012 was perhaps the most widely circulated piece.

34. See Guerrero 2013.

35. Simpson writes extensively of the conflation of indigeneity with criminality and "lawless savagery" in the context of settler colonialism in part because settler law is intrinsically "precarious and fragile" because of its violent imposition and legacy. For this reason, settler law is determined to extinguish presettler indigeneity at all costs (2014, 2016).

36. The new governor, Alejandro Murat Hinojosa, is the son of former governor José Murat Casab, who was instrumental in supporting the early stages of Oaxacan wind development.

4. DISTRITO FEDERAL

1. See the *Ecologics* volume.

2. Raúl Salinas de Gortari is the elder brother of former Mexican president Carlos Salinas de Gortari (1988–94), and he is infamous in Mexico for political corruption, money laundering, and drug trafficking. Raúl Salinas spent ten years in prison for homicide before being acquitted in 2005. In July 2013 the last remaining charge of "unlawful enrichment" against him was dismissed, allowing him to recover a substantial portion of his former fortune. This landed him on *Forbes*'s list of "The 10 Most Corrupt Mexicans of 2013."

3. The energy-reform process in 2013 did not fail to offer a spectacular watershed in the history of the Mexican petrostate. For a relatively balanced accounting of the reform process and its outcomes, see Columbia Center on Global Energy Policy, https://energypolicy.columbia.edu/research/global-energy-dialogue/mexican-energy-reform-prospects-and-challenges.

4. Webber 2014.

5. See, e.g., Appel, Mason and Watts 2015; Behrends, Reyna, and Schlee 2011; McNeish and Logan 2012; Reyna and Behrends 2008; Sawyer 2004, 2007; Sawyer and Gomez 2012.

6. Doug Rogers (2015) analyzes similar dynamics in his study of Lukoil's culture industry in Russia.

7. See Campbell (2014) on narco-propaganda and van Dun (2014) on narco-sovereignty, for example. See also Kruijt 2012.

8. The particular story in this case was SENER's claim that a refinery would be fully operational in 2015 when the building had scarcely been started in 2013. Laurence explained to us that even once completed, a refinery takes three to five years to "fine tune" in order to ensure its stable operation.

9. In an interview with us in 2014, former president Calderón admitted that the figure had been as high as 43 percent during his administration.

10. This is precisely what happened when international oil prices fell sharply in the fourth quarter of 2014, from a high of $112 per barrel in June to $62 per barrel in December. A series of political and financial crises followed for Peña Nieto's administration in 2015 and 2016, including a requirement that they remove subsidies on the domestic use of gas, leaving him with a 17 percent approval rating in early 2017. See, e.g., Tillman 2017.

11. This claim—that Mexico offered extraordinary subsidies to its poorest citizens, offset in part by higher electricity prices for industry—was one we heard relatively often, including from CFE employees themselves. It has a kernel of truth in that Mexico has a tiered pricing system for electricity usage that favors less use of electricity in nominal terms. And industrial electricity tariffs are about 70 percent higher in Mexico than they are, for example, across the border in the United States. A World Bank study (Komives et al. 2009) concluded, however, that "Mexico's electricity tariff structures are among the most complex in the world, by design and by natural accretion" (45), and that its subsidies are also among the world's largest (an estimated $9 billion in 2006), with two-thirds of those subsidies going to residential consumers (vii) while "disproportionately benefit[ing] large-volume consumers" (20).

12. Wionczek 1965, 542.

13. The years when Porfirio Díaz and his allies ruled, 1876–1911, are generally referred to as the Porfiriato.

14. The last chairman of MexLight, Maxwell Taylor, commented in his memoir, "MexLight was about as international as a business enterprise could be, having its headquarters in Toronto, its principal stockholders in Belgium, the source of most of its capital in Wall Street and the market for its product in Mexico. Its staff was made up largely of Mexicans but with a considerable sprinkling of Americans and Europeans among the officers and technicians. Like most utilities its growth was fettered by the power rates which in our case were set by the Mexican government; the latter was not inclined toward generosity to a corporation which was popularly regarded as a symbol of Yankee industrial penetration. Because of the low rate schedule, the company was having great difficulty in expanding its capacity to meet the population and industrial growth in its area of responsibility, and consequently it often incurred the ire of its customers for inadequate service" (M. Taylor 1972); see also Hausman and Neufeld 1997; Niblo 1999.

15. The aforementioned World Bank report (Komives et al. 2009) estimates Luz y Fuera's distribution losses as "very high, exceeding 30%" (13).

16. See Belmont 2012 for a close analysis of the conflict between the SME and the government. For international coverage of the Luz y Fuerza closure, see de Córdoba 2009; Lacey 2009.

17. As of the end of 2015, the percentage of electricity on the Mexican national grid produced by CFE had dropped to 57.2 percent (Melgar, Díaz de Leon, and Luque 2015, 26).

18. The energy reform incentivized independent power production as well as private sector investment. In 2015, SENER proudly projected $116 billion of new investment opportunities in generation, transmission, and distribution across the national grid between 2015 and 2030 (Melgar, Díaz de Leon, and Luque 2015, 28). The SENER plan has a strong emphasis on natural gas and renewable energy sources with a target of 50 percent clean electricity by 2050. Speaking at Rice University in September 2016, Mexican energy secretary Pedro Joaquin Coldwell stated that Mexico has already received $22.4 billion in investment commitments from fifty-nine private companies since the energy-reform process began. However, $19.5 billion of that investment has been focused on oil and gas exploration and pipelines (Hunn 2016).

19. Iliff 2015.

20. SENER's current five-year plan (2014–19) calls for ten new strategic gas pipelines and seven interconnections with the United States (Melgar, Díaz de Leon, and Luque 2015, 19).

21. Arguments concerning baseload have recently become a front line in the struggle for decarbonizing energy in the United States as well with fossil fuel lobbyists and political actors like US energy secretary Rick Perry declaring their commitment to defend baseload with others, notably renewable energy financier and advocate Jigar Shah, declaring, "There is no such thing as baseload." See Shah 2015; Joyce 2016.

22. Boyer 2015. Cf. Bakke 2016; Hughes 1983.

23. Özden-Schilling (2015) argues, however, that the unique materiality of electricity challenges conventional political-economic models of commodities and markets.

24. Popular imaginaries of electricity, including those expressed to us by some CFE employees, tend to conceptualize electricity in terms of "flow," as though electrons flow through wires like water through pipes, surfacing the hydropolitics of water and sewage that have historically informed public understanding of other critical infrastructures of modernity. Other electric administrators and engineers were quick to note that the more accurate analogy is to think about electricity supply in terms of "load," hence "baseload" as an "essential, constant supply."

25. See, e.g., NERC 2014.

26. The Kyoto Protocol is an international treaty based on the 1992 United Nations Framework Convention on Climate Change (UNFCCC) in which the signatories have committed to reduce greenhouse gas emissions.

27. See Comisión Reguladora de Energía 2012.

28. As of the time of this writing, Peña Nieto's government has maintained Calderón's commitment to 35 percent electricity from non-fossil-fueled sources by 2024.

29. In 2016, as part of the energy reform process, management of the grid was transferred from CFE to a former subunit, CENACE, Centro Nacional de Control de Energía, which is now organized as the independent system operator for the whole national power grid. The commission's power generation assets were meanwhile divided into four subsidiary generation companies, which will sell electricity into a new wholesale market. See Manzagol and Hodge 2016.

30. Calderón changed the step-like tariff system where a single extra kilowatt-hour would boost people into a higher bracket and replaced it with a line-like tariff system with fewer "discontinuities." Villagómez explained how, in old stepped system, at five hundred kilowatt-hours of consumption for two months, an electricity bill would be around nine hundred pesos but at 501 kilowatt-hours, it would jump to 2,500 pesos. This was again a "political" rather than "technical" decision, according to Villagómez, but "since most people don't follow their electricity usage very carefully, the system was not clear to people and created a lot of anger that CFE seemed to be setting arbitrary prices."

31. We later learned that the governor who approached her was Diódoro Carrasco.

32. Including the Ministry of the Interior (SEGOB), the Environment Ministry (SEMARNAT), and the National Commission for the Development of Indigenous Peoples (CDI).

33. See Secretaría de Energía 2015.

34. See Anaya 2015.

35. See chapter 3 of the *Ecologics* volume.

36. See Derrida 1976.

37. This "ecological" dimension of enablement is, I would submit, an enduring lesson of the anthropology of infrastructure, something that numerous studies have explored and revealed in their own fashion. See, e.g., Anand 2012; Appel, Anand, and Gupta 2015; Barry 2013; Bowker 2010; Harvey, Bruun Jensen, and Morita 2017; Harvey and Knox 2015; Larkin 2013; Mitchell 2011; Star and Ruhleder 1996; Star 1999; von Schnitzler 2013.

5. GUIDXIGUIE' (JUCHITÁN DE ZARAGOZA)

1. For a study of the APPJ, see Ortiz Rubén et al. 2014.

2. Martínez López (1966) offers a detailed account of the battle itself.

3. The Gas Natural Fenosa park eventually became operational despite the resistance. To date, the only wind park project that has been successfully blocked by the antieólico resistance has been the Mareña Renovables project in San Dionisio del Mar (see the *Ecologics* volume of the duograph).

4. Radio Totopo is a locally celebrated station that has been broadcasting in diidxazá (Zapotec) from the Séptima Sección of Juchitán since 2005 (Nava Morales 2015). Since 2008, the station has been actively involved in organizing and publicizing resistance against wind development in the isthmus. See, e.g., "Pronunciamiento de la asamblea de radios libres y comunitarias de oaxaca," Radio Totopo, August 31, 2008, http://alimentandolaresistenciaradiototopo.blogspot.com. On the importance of community and indigenous radio in Oaxaca, see Stephen 2012. On the politics of indigenous linguistic revival in Mexico, see Faudree 2013, 2015.

5. #YoSoy132 was a mass student and youth protest movement in Mexico that developed in response to the 2012 presidential campaign and, in particular, in opposition to the return to power of the PRI party and their controversial candidate Enrique Peña Nieto (see Favela 2015). #YoSoy132 made exceptional use of social media (Bacallao-

Pino 2016; Treré 2015), which helped them to attract international visibility and support even though they were, strictly speaking, unsuccessful in their effort to prevent a PRI victory.

6. This is one of the more legendary tunes within what Conant has termed the "poetics of resistance" within zapatismo (2010).

7. Conquistadors Pedro de Alvarado and Hernán Cortés were redheaded according to legend.

8. This performance is one that, Audra Simpson reminds us (2014), has been solicited in no small part by the settler empire's effort to conjure and stabilize the imagination of a militant indigenous savage other, which then operates as the alibi for further violent intervention. Simpson has theorized this relation with respect to the Mohawk and questions of "how one is to define a citizenship for one's own people, according to one's political traditions while operating in the teeth of Empire, in the face of state aggression."

9. For further glimpses into the complex ecology of power in Juchitán see, e.g., Bailón and Zermeño 1987; Campbell 1990; Campbell et al. 1993; Kraemer Bayer 2008; Rubin 1997.

10. On these dynamics in Mexico, see especially Liffman 2011; Muehlmann 2013; Rodríguez 2002; Taylor 2009. More broadly, I have in mind the works of scholars such as Coulthard 2007, 2014; Kauanui 2008; Moreton-Robinson 2015; Povinelli 2002; Simpson 2014; Veracini 2015.

11. Povinelli 2002.

12. For deeper insight into mestizaje nationalism, see Alonso 2004; Bartra 2002; Loewe 2011; Lomnitz 2001. My analytic strategy in this chapter intends no disrespect to the rich tradition of anthropological analysis of and cultural commentary on mestizaje relations and thinking in Mexico. What I am trying to do is show that recent critical research on settler liberalism—including especially settler statecraft and indigenous sovereignty—has useful insights to share in the context of understanding Juchitán, whose citizens commonly understand themselves to live outside the sphere of mestizaje nationalism in a zone of enduring, if contested, indigenous autonomy on the frontier of Mexican law and state violence.

13. What the 1917 Mexican Constitution's famous Article 27 granted was not sovereignty to indigenous peoples but rather usufructuary rights to "centers of population which, by law or in fact, possess a communal status . . . to enjoy common possession of the lands, forests, and waters belonging to them or which have been or may be restored to them." Meanwhile, "ownership of the lands and waters within the boundaries of the national territory is vested originally in the Nation, which has had, and has, the right to transmit title thereof to private persons, thereby constituting private property." Bienes comunales and bienes ejidales thus become specific forms of private property awarded by the nation—itself conceived as an indivisible sovereign subject—to groups, including indigenous peoples, over which it continues to exert a right of expropriation "for reasons of public use and subject to payment of indemnity." See "1917 Constitution of Mexico," Latin American Studies, http://www.latinamericanstudies.org/mexico/1917-Constitution.htm; cf. Simpson 2014.

14. In a fascinating article, Kourí (2002) has further argued that the objective in restoring land and some political autonomy to indigenous pueblos was not multicultural freedom but rather a Comtean-Spencerian logic of creating the most efficient vehicle for the peaceful evolution and eventual assimilation of indigenous peoples into the Mexican nation.

15. I mean this in Simpson's (2016) sense of "refusal" as a fundamental unwillingness to accept the sovereign claims of settler governance: "Refusal holds on to a truth, structures this truth as stance through time, as its own structure and comingling with the force of presumed and inevitable disappearance and operates as the revenge of consent—the consent to these conditions, to the interpretation that this was fair, and the ongoing sense that this is all over with."

16. According to Broadwell (2015), "The historical circumstances that led to the movement of some Valley Zapotecs toward the Isthmus are complex and the subject of ongoing research among historians, but the emerging view . . . seems to be that the earlier Zapotec kingdom was based in Zaachila in the Valley of Oaxaca. It began to expand toward the Pacific Coast and the Isthmus of Tehuantepec in the fourteenth century. For about 150 years, there was both a Valley and an Isthmus center of power, with tensions between the two, and movements of populations from one center to the other." See also King 2012.

17. See Saynes-Vásquez 2002, 40; see also Tutino 1993; Zeitlin 1994.

18. See, e.g., King 2012.

19. See Oudijk 2000; Zeitlin 1994.

20. See Gutiérrez Brockington 1989.

21. Taussig (1987) has documented the cultural legacy of this brutal period unforgettably. See also Patel and Moore 2017.

22. Tutino 1980.

23. Tutino 1980; translated by the author.

24. The repartimiento system flourished in Mexico between 1550s and 1630s, guaranteeing indigenous labor for the Spaniards during a time of massive population loss while familiarizing Indians with temporary wage labor and patterns of hacienda production. "Like other colonial corveés, therefore, the *repartimiento* system indirectly helped pave the way for freer forms of proletarian labour" (Knight 2002, 83). Taylor supports the standard critical account of repartimiento in Oaxaca but notes that because of labor shortages and the difficulty of contracting labor, Oaxacan peasants were actually paid relatively well on Spanish colonial estates in the seventeenth and eighteenth centuries although instances of coercion were common (W. Taylor 1972). With the help of neoclassical economic theory, Baskes (2000) even reimagines Oaxacan repartimiento as a mostly benign credit system (cf. Ruiz Medrano 2005).

25. Anderson 1983; see also Sánchez Santiró 2001.

26. Zeitlin writes, more specifically, "Virtually the entire province was kept busy meeting diverse labor demands for Avellán's gain: salt-making, fishing, preparing deer hides, hunting and collecting rabbit wool for hats. He forced communities to sell him vanilla, which at least was produced locally, but paid them only half the going price.

Worst of all, however, was the cotton cloth quota, which no other *alcade mayor* had attempted to institute in recent memory. Since cotton was not produced locally at the time, people had to journey all the way to Chiapas" (2005, 184).

27. Zeitlin 2005, 169.

28. See Rojinsky 2008 for a subtle analysis of the case.

29. Zeitlin 2005, 213ff.

30. See, e.g., Díaz-Polanco and Burguete 1989.

31. See de la Cruz 1983; Chassen-López 2004.

32. Chassen-López 2004, 321.

33. Rippy (1920) explains in more detail the complexities of the political situation, including both the fallout from the US-Mexican War and the jealousies among those advocating for alternative transoceanic routes in Panama and Nicaragua.

34. Chassen-López 2004, 326.

35. Ristow 2008, 105.

36. Díaz's circle of positivist technocrats whose combined use of infrastructure and foreign capital to modernize Mexico remains controversial to this day. See, e.g., Hale 1991; Katz 2004; Raat 1975; Vélazquez Becerril 2010.

37. Chassen-López 2004, 328–29.

38. This was known subsequently as the Chegomista Rebellion; see de la Cruz 1990; Purnell 2005; Chassen-López 2009, 30–33.

39. Ristow 2008, 170.

40. Ristow 2008, 106.

41. Ristow 2008, 169.

42. Smith 2012, 119; see also Rubin 1997, 53.

43. See, e.g., de la Cruz 1993 on the "pacification" of post-Revolutionary Mexico.

44. Smith 2012, 121.

45. See the *Ecologics* volume.

46. On the origins and rise to power of the COCEI, see López Villalobos 2004.

47. See, e.g., Bailón Corres and Zermeño 1987; Campbell 1990.

48. On global indigenist politics, see Merlan 2009; on the rise of neoanarchism, see Graeber 2002; Boyer 2013b. On contemporary indigenous responses to settler energy infrastructure, see "#StandingRockSyllabus," NYC Stands with Standing Rock Collective, September 5, 2016, https://nycstandswithstandingrock.wordpress.com /standingrocksyllabus/.

49. See, e.g., Terán 2012.

50. Zanates are *Quiscalus mexicanus*, great-tailed grackles.

51. Sadly, David passed away during the writing of the duograph.

52. Throughout his mayorship, Vicente was accused by both supporters and critics of wind power development of not doing enough to support their cause, evidence perhaps that he was trying to remain neutral.

53. "Bettina Cruz: Mexican Human Rights Defender Working on Business and Human Rights," International Service for Human Rights, January 6, 2015, http://www .ishr.ch/news/bettina-cruz-mexican-human-rights-defender-working-business-and -human-rights.

54. See, e.g., "Oaxaca: Policía de Juchitán balacea a compañeros de la Asamblea Comunitaria de Alvaron Obregón," Centro de Medios Libres, May 14, 2016, https://www.centrodemedioslibres.org/2016/05/14/oaxaca-policia-de-juchitan-balacea-a-companeros-de-la-asamblea-comunitaria-de-alvaron-obregon/.

55. See the *Ecologics* volume.

56. "Frontier" is meant here in the sense developed by Tsing 2015; Patel and Moore 2017.

References

Abélès, Marc, ed. 2011. *Des anthropologues a l'OMC*. Paris: CNRS.

Abrams, Philip. 1988. "Notes on the Difficulty of Studying the State." *Journal of Historical Sociology* 1: 58–89.

Adams, Richard Newbold. 1975. *Energy and Structure: A Theory of Social Power*. Austin: University of Texas Press.

Adams, Richard Newbold. 1978. "Man, Energy, and Anthropology: I Can Feel the Heat, but Where's the Light?" *American Anthropologist* 80: 297–309.

Adler-Lomnitz, Larissa, Rodrigo Salazar-Elena, and Ilya Adler. 2010. *Symbolism and Ritual in a One-Party Regime*. Translated by Susanne A. Wagner. Tucson: University of Arizona Press.

Agamben, Giorgio. 1998. *Homo Sacer: Sovereign Power and Bare Life*. Stanford, CA: Stanford University Press.

Ahmed, Sara. 2014. *Willful Subjects*. Durham, NC: Duke University Press.

Aiello, José L., Jorge I. Valencia, Enrique Caldera Muñoz, and Vicente L. Gómez. 1983. *Atlas Eolico Preliminar de America Latina y el Caribe*. Quito: OLADE.

Aitken, Rob, Nikki Craske, Gareth A. Jones, and David E. Stansfield, eds. 1996. *Dismantling the Mexican State?* London: Macmillan.

Almazan, Jose Antonio. 1994. *Electricidad: Una nacionalización inconclusa*. Mexico City: Multiediciones California.

Alonso, Ana Maria. 2004. "Conforming Disconformity: 'Mestizaje,' Hybridity, and the Aesthetics of Mexican Nationalism." *Cultural Anthropology* 19 (4): 459–90.

Anand, Nikhil. 2012. "Pressure: The Politechnics of Water Supply in Mumbai." *Cultural Anthropology* 26 (4): 542–64.

Anaya, S. James. 2015. "Observaciones del profesors: James Anaya sobre la consulta en el contexto del proyecto energía eólica del sur en Juchitán de Zaragoza." February 23. https://consultaindigenajuchitan.files.wordpress.com/2015 /01/juchitan-observaciones-anaya.pdf.

Anderson, Benedict. 1983. *Imagined Communities*. London: Verso.

Anusas, Mike, and Tim Ingold. 2015. "The Charge against Electricity." *Cultural Anthropology* 30 (4): 540–54.

Apffel-Marglin, Frédrique, and Stephen A. Marglin. 1990. *Dominating Knowledge: Development, Culture and Resistance*. Oxford: Clarendon.

Appel, Hannah, Nikhil Anand, and Akhil Gupta. 2015. "Introduction: The Infrastructure Toolbox." *Cultural Anthropology* website, September 24. https://culanth.org/fieldsights/714-introduction-the-infrastructure-toolbox.

Appel, Hannah, Arthur Mason, and Michael Watts, eds. 2015. *Subterranean Estates: Life Worlds of Oil and Gas*. Ithaca, NY: Cornell University Press.

Bacallao-Pino, Lázaro M. 2016. "Radical Political Communication and Social Media: The Case of the Mexican #YoSoy132." In *(R)evolutionizing Political Communication through Social Media*, edited by Tomaz Dezelan and Igor Vobic, 56–74. Hershey, PA: IGI Global.

Bailón Corres, Moisés J., and Sergio Zermeño. 1987. *Juchitán: Límites de una Experiencia Democrática*. Mexico City: Instituto de Investigaciones Sociales.

Bakke, Gretchen. 2016. *The Grid: The Fraying Wires between Americans and Our Energy Future*. New York: Bloomsbury.

Barad, Karen. 2003. "Posthumanist Performativity: Toward an Understanding of How Matter Comes to Matter." *Signs* 28 (3): 801–31.

Barry, Andrew. 2013. *Material Politics: Disputes along the Pipeline*. New York: Wiley.

Bartra, Roger. 2002. *Blood, Ink and Culture: Miseries and Splendors of the Post-Mexican Condition*. Durham, NC: Duke University Press.

Bartra, Roger, and Eugenia Huerta. 1978. *Caciquismo y Poder Politico en el México Rural*. Mexico City: Siglo Veintiuno Editores.

Baskes, Jeremy. 2000. *Indians, Merchants and Markets: A Reinterpretation of the Repartimiento and Spanish-Indian Economic Relations in Colonial Oaxaca, 1750–1821*. Stanford, CA: Stanford University Press.

Beck, Stefan. 2007. "Medicalizing Culture(s) or Culturalizing Medicine(s)?" In *Medicine as Culture: Instrumental Practices, Technoscientific Knowledge, and New Modes of Life*, edited by Regula Valerie Burri and Joseph Dumit, 17–33. London: Routledge.

Behrends, Andrea, Stephen Reyna, and Guenther Schlee, eds. 2011. *Crude Domination: The Anthropology of Oil*. New York: Berghahn.

Belmont, Edgar. 2012. "Luz y Fuerza del Centro: ejes del conflicto entre el Sindicato Mexicano de Electricistas y el Gobierno Federal." *Revista Estudios Sociológicos* XXX (89): 331–65.

Bennett, Jane. 2009. *Vibrant Matter: A Political Ecology of Things*. Durham, NC: Duke University Press.

Benton, Allyson. 2011. "The Origins of Mexico's Municipal Usos y Costumbres Regimes: Supporting Local Political Participation or Local Authoritarian Control?" Online publication. http://libreriacide.com/librospdf/DTEP-226.pdf.

Bernal, Victoria. 2013. "Please Forget Democracy and Justice: Eritrean Politics and the Powers of Humor." *American Ethnologist* 40 (2): 300–309.

Biehl, João. 2007. *Will to Live: AIDS Therapies and the Politics of Survival*. Princeton, NJ: Princeton University Press.

Binford, Leigh. 1985. "Political Conflict and Land Tenure in the Isthmus of Tehuantepec." *Journal of Latin American Studies* 17 (1): 177–200.

Bini, Elisabetta, and Giuliano Garavini, eds. 2016. *Oil Shock: The 1973 Crisis and Its Economic Legacy*. London: Tauris.

Bobick, Michael S. 2012. "Performative Sovereignty: State Formation in the Transnistrian Republic." PhD diss., Cornell University.

Bobick, Michael S. 2014. "Separatism Redux: Crimea, Transnistria, and Eurasia's *De Facto* States." *Anthropology Today* 30 (3): 3–8.

Bogost, Ian. 2012. *Alien Phenomenology, or What It's Like to Be a Thing*. Minneapolis: University of Minnesota Press.

Borja Díaz, Marco Antonio R., Oscar A. Jaramillo Salgado, and Fernando Mimiaga Sosa. 2005. *Primer Documento del Proyecto Eoloeléctrico del Corredor Eólico del Istmo de Tehuantepec*. Mexico City: Instituto de Investigaciones Eléctricas.

Bornstein, Erica. 2005. *The Spirit of Development: Protestant NGOs, Morality and Economics in Zimbabwe*. Palo Alto, CA: Stanford University Press.

Bourdieu, Pierre. 1991. *Language and Symbolic Power*. Translated by Peter Collier. Cambridge: Polity.

Bowker, Geoffrey C. 2010. "Towards Information Infrastructure Studies: Ways of Knowing in a Networked Environment." In *The International Handbook of Internet Research*, edited by Jeremy Hunsinger, Lisbeth Klastrup, and Matthew Allen, 97–117. New York: Springer.

Boyer, Dominic. 2003. "Censorship as a Vocation: The Institutions, Practices, and Cultural Logic of Media Control in the German Democratic Republic." *Comparative Studies in Society and History* 45 (3): 511–45.

Boyer, Dominic. 2005. *Spirit and System: Media, Intellectuals and the Dialectic in Modern German Culture*. Chicago: University of Chicago Press.

Boyer, Dominic. 2010. "On the Ethics and Practice of Contemporary Social Theory: From Crisis Talk to Multiattentional Method." *Dialectical Anthropology* 34 (3): 305–24.

Boyer, Dominic. 2013a. *The Life Informatic: Newsmaking in the Digital Era*. Ithaca, NY: Cornell University Press.

Boyer, Dominic. 2013b. "Simply the Best: Parody and Political Sincerity in Iceland." *American Ethnologist* 40 (2): 276 87.

Boyer, Dominic. 2014. "Energopower: An Introduction." *Anthropological Quarterly* 87 (2): 309–34.

Boyer, Dominic. 2015. "Anthropology Electric." *Cultural Anthropology* 30 (4): 531–39.

Boyer, Dominic. 2016. "Revolutionary Infrastructure." In *Infrastructures and Social Complexity: A Companion*, edited by Penelope Harvey, Casper Bruun Jensen, and Atsuro Morita, 174–86. London: Routledge.

Boyer, Dominic, James D. Faubion, and George E. Marcus, eds. 2015. *Theory Can Be More Than It Used to Be: Learning Anthropology's Method in a Time of Transition*. Ithaca, NY: Cornell University Press.

Boyer, Dominic, and Cymene Howe. 2015. "Portable Analytics and Lateral Theory." In *Theory Can Be More Than It Used to Be: Learning Anthropology's Method in a Time of Transition*, edited by Dominic Boyer, James D. Faubion, and George E. Marcus, 15–38. Ithaca, NY: Cornell University Press.

Boyer, Dominic, and George Marcus. Forthcoming. Introduction to *Collaborative Anthropology Today*. Ithaca, NY: Cornell University Press.

Boyer, Dominic, and Timothy Morton. 2016. "Hyposubjects." Theorizing the Contemporary, *Cultural Anthropology* website. January 21. https://culanth.org /fieldsights/798-hyposubjects.

Boyer, Dominic, and Alexei Yurchak. 2010. "American Stiob: Or, What Late-Socialist Aesthetics of Parody Reveal about Contemporary Political Culture in the West." *Cultural Anthropology* 25 (2): 179–221.

Breceda, Miguel. 2000. "Debate on Reform of the Electricity Sector in Mexico: Report on Its Background, Current Status and Outlook." Report prepared for the North American Commission for Environmental Cooperation. http://www3.cec.org /islandora/en/item/1611-debate-reform-electricity-sector-in-mexico-en.pdf.

Breglia, Lisa. 2013. *Living with Oil: Promises, Peaks, and Declines on Mexico's Gulf Coast*. Austin: University of Texas Press.

Briggs, Charles L. 2005. "Communicability, Racial Discourse and Disease." *Annual Review of Anthropology* 34: 269–91.

Briggs, Charles L., and Mark Nichter. 2009. "Biocommunicability and the Biopolitics of Pandemic Threats." *Medical Anthropology* 28 (3): 189–98.

Broadwell, George Aaron. 2015. "The Historical Development of the Progressive Aspect in Central Zapotec." *International Journal of American Linguistics* 81 (2): 151–85.

Brown, Jennifer. 2004. *Ejidos and Comunidades in Oaxaca, Mexico: Impact of the 1992 Reforms*. Rural Development Institute Reports on Foreign Aid and Development 120. Seattle: RDI.

Brown, Wendy. 2015. *Undoing the Demos: Neoliberalism's Stealth Revolution*. Cambridge, MA: MIT Press.

Brulotte, Ronda. 2009. "'Yo soy nativo de aquí': The Ambiguities of Race and Indigeneity in Oaxacan Craft Tourism." *Journal of Latin American and Caribbean Anthropology* 14 (2): 457–82.

Burke, John Francis. 2002. *Mestizo Democracy: The Politics of Crossing Borders*. College Station: Texas A&M University Press.

Burkett, Paul, and John Bellamy Foster. 2006. "Metabolism, Energy and Entropy in Marx's Critique of Political Economy: Beyond the Podolinsky Myth." *Theory and Society* 35: 109–56.

Cabrero Mendoza, Enrique. 2005. "Federalism and Decentralization in Mexico: The Dilemmas of Municipalities." Unpublished paper.

Cabrero Mendoza, Enrique. 2007. "Government Decentralization and Decentral- ized Governance in Latin America: The Silent Revolution at the Local Level?" In *Decentralizing Governance: Emerging Concepts and Practices*, edited by G. Shabbir Cheema and Dennis A. Rondinelli, 156–69. Washington, DC: Brookings Institution.

Cahn, Peter S. 2008. "Consuming Class: Multilevel Marketers in Neoliberal Mexico." *Cultural Anthropology* 23 (3): 429–52.

Caldera Muñoz, Enrique, et al. 1980. "Estudio preliminar y potencial de La Ventosa, Oax., para el aprovechamiento de la energía eólica." *Boletín IIE* 4 (8/9): 46–57.

Caldera Muñoz, Enrique, and R. Saldaña Flores. 1986. *Evaluación Preliminar del Potencial de Generación Eléctrica en la Zona de La Ventosa*. Oaxaca: Instituto de Investigaciones Eléctricas, Informe Técnico FE/01/14/2063/1-01/P.

Campbell, Howard. 1990. "Juchitán: The Politics of Cultural Revivalism in an Isthmus Zapotec Community." *Latin American Anthropology Review* 2 (2): 47–55.

Campbell, Howard. 2014. "Narco-Propaganda in the Mexican 'Drug War': An Anthropological Perspective." *Latin American Perspectives* 41 (2): 60–77.

Campbell, Howard, Leigh Binford, Miguel Bartolomé, and Alicia Barabas, eds. 1993. *Zapotec Struggles: Histories, Politics, and Representations from Juchitán, Oaxaca*. Washington, DC: Smithsonian Institution Press.

Castellanos, Bianet M. 2010. "Don Teo's Expulsion: Property Regimes, Moral Economies, and Ejido Reform." *Journal of Latin American and Caribbean Anthropology* 15 (1): 144–69.

Chalfin, Brenda. 2010. *Neoliberal Frontiers: An Ethnography of Sovereignty in West Africa*. Chicago: University of Chicago Press.

Chance, Kerry Ryan. 2015. "'Where There Is Fire, There Is Smoke': Ungovernability and Material Life in Urban South Africa." *Cultural Anthropology* 30 (3): 394–423.

Chassen-López, Francie R. 2004. *From Liberal to Revolutionary Oaxaca: The View from the South, Mexico 1867–1911*. State College: Pennsylvania State University Press.

Chassen-López, Francie R. 2009. "Benito Juaréz Maza of Oaxaca: A Revolutionary Governor?" In *State Governors in the Mexican Revolution, 1910–1952: Portraits in Conflict*, edited by Jürgen Buchenau and William H. Beezley, 19–42. New York: Rowman & Littlefield.

Chavez, Xóchitl C. 2013. "Migrating Performative Traditions: The Guelaguetza Festival in Oaxacalifornia." PhD diss., University of California Santa Cruz.

Christensen, Benny. 2013. "History of Danish Wind Power." In *Wind Power for the World: The Rise of Modern Wind Energy*, edited by Preben Maegaard, Anna Krenz, and Wolfgang Palz, 33–92. Boca Raton, FL: CRC.

Clarke, Colin. 2000. *Class, Ethnicity and Community in Southern Mexico: Oaxaca's Peasantries*. Oxford: Oxford University Press.

Coerschulte, Tatjana. 2014. "So sieht eine Geldmaschine aus: Üppige Einkünfte dank Windkraft." HNA. September 3. http://www.hna.de/politik/pachten-windrad -standorte-erzielen-grundbesitzer-ueppige-einkuenfte-3829406.html.

Cohen, Jeffrey H. 2004. *The Culture of Migration in Southern Mexico*. Austin: University of Texas Press.

Cohen, Lawrence. 2005. "Operability, Bioavailability, and Exception." In *Global Assemblages: Technology, Politics, and Ethics as Anthropological Problems*, edited by Aihwa Ong and Stephen J. Collier, 79–90. Malden, MA: Blackwell.

Colebrook, Claire. 2017. "We Have Always Been Post-Anthropocene: The Anthropocene Counter-Factual." In *Anthropocene Feminisms*, edited by Richard Grusin, 1–20. Minneapolis: University of Minnesota Press.

Comaroff, Jean, and John L. Comaroff. 1999. "Occult Economies and the Violence of Abstraction: Notes from the South African Postcolony." *American Ethnologist* 26 (2): 279–303.

Comaroff, Jean, and John L. Comaroff, eds. 2001. *Millennial Capitalism and the Culture of Neoliberalism*. Durham, NC: Duke University Press.

Comisión Federal de Electricidad. 2012. *Programa de obras e inversiones del sector eléctrico 2012–2026*. Mexico City: Comisión Federal de Electricidad.

Comisión Reguladora de Energía. 2012. *Memoria descriptiva: Temporadas abiertas de reserva de capacidad de transmisión y transformación*. October. Mexico City: Comisión Reguladora de Energía.

Conant, Jeff. 2010. *A Poetics of Resistance: The Revolutionary Public Relations of the Zapatista Insurgency*. Oakland, CA: AK.

Cornelius, Wayne A., and David Myhre, eds. 1998. *The Transformation of Rural Mexico: Reforming the Ejido Sector*. La Jolla: Center for US-Mexican Studies, University of California San Diego.

Coronil, Fernando. 1997. *The Magical State: Nature, Money, and Modernity in Venezuela*. Chicago: University of Chicago Press.

Coulthard, Glen. 2007. "Subjects of Empire: Indigenous Peoples and the 'Politics of Recognition' in Canada." *Contemporary Political Theory* 6: 437–60.

Coulthard, Glen. 2014. *Red Skin, White Masks: Rejecting the Colonial Politics of Recognition*. Minneapolis: University of Minnesota Press.

Covarrubias, Miguel. 1946. *Mexico South: The Isthmus of Tehuantepec*. New York: Knopf.

Crate, Susan. 2011. "Climate and Culture: Anthropology in the Era of Contemporary Climate Change." *Annual Review of Anthropology* 40: 175–94.

Crate, Susan, and Mark Nuttall, eds. 2009. *Anthropology and Climate Change*. Walnut Creek, CA: Left Coast.

Crewe, Emma, and Richard Axelby. 2013. *Anthropology and Development: Culture, Morality and Politics in a Globalised World*. Cambridge: Cambridge University Press.

Daggett, Cara. 2019. *The Birth of Energy: Fossil Fuels, Thermodynamics and the Politics of Work*. Durham, NC: Duke University Press.

Das, Veena, and Deborah Poole. 2004. "State and Its Margins: Comparative Ethnographies." In *Anthropology in the Margins of the State*, edited by Veena Das and Deborah Poole, 3–34. Santa Fe, NM: School of American Research Press.

Davis, Mike. 2010. "Who Will Build the Ark?" *New Left Review* 61 (January/February): 29–46.

De Córdoba, José. 2009. "Mexico Power Takeover Creates Sparks." *Wall Street Journal*, October 12. http://www.wsj.com/articles/SB125530734212079359.

Degani, Michael. 2017. "Modal Reasoning in Dar es Salaam's Power Network." *American Ethnologist* 44 (2): 300–314.

De Ita, Ana. 2003. *Mexico: The Impacts of Demarcation and Titling by PROCEDE on Agrarian Conflicts and Land Concentration*. Mexico: Centro de Estudios para el Cambio en el Campo Mexicano, Land Research Action Network.

De la Cruz, Victor. 1983. *La Rebelión de Che Gorio Melendre*. Juchitán: Ayuntamiento Popular de Juchitán.

De la Cruz, Victor. 1990. "Che Gomez y la rebelión de Juchitán: 1911." In *Lecturas Historicas de Estado de Oaxaca*, vol. 4, Siglo XIX, edited by María de los Angeles Romero Frizzi, 247–71. Mexico City: INAH.

De la Cruz, Victor. 1993. *El general Charis y la pacificatión del México posrevolucionario*. Mexico City: CIESAS.

Deleuze, Gilles. 1995. *Negotiations*. New York: Columbia University Press.

Derrida, Jacques. 1976. *Of Grammatology*. Translated by Gayatri Chakravorty Spivak. Baltimore: Johns Hopkins University Press.

Díaz-Polanco, Héctor, and Araceli Burguete. 1989. "Sociedad Colonial y Rebelión Indígena en el Istmo de Tehuantepec." *Boletín de Antropología Americana* 20 (December): 99–124.

Dietrich, Christopher R. W. 2008. "The Permanence of Power: The Energy Crisis, Sovereign Debt, and the Rise of American Neoliberal Diplomacy, 1967–1976." PhD diss., University of Texas at Austin.

Dove, Michael, ed. 2014. *The Anthropology of Climate Change: A Historical Reader*. New York: Wiley.

Duménil, Gérard, and Dominique Lévy. 2011. *The Crisis of Neoliberalism*. Cambridge, MA: Harvard University Press.

Edelman, Marc, and Angélique Haugerud. 2005. "Introduction: The Anthropology of Development and Globalization." In *The Anthropology of Development and Globalization: From Classical Political Economy to Contemporary Neoliberalism*, edited by Marc Edelman and Angélique Haugerud, 1–75. Malden, MA: Blackwell.

El Khachab, Chihab. 2016. "Living in Darkness: Internet Humour and the Politics of Egypt's Electricity Infrastructure." *Anthropology Today* 32 (4): 21–24.

Elliott, Dennis, Marc Schwartz, Steve Haymes, Donna Heimiller, and R. George. 2003. *Wind Energy Resource Atlas of Oaxaca*. Oak Ridge, TN: NREL, US Department of Energy.

Elyachar, Julia. 2005. *Markets of Dispossession: NGOs, Economic Development, and the State in Cairo*. Durham, NC: Duke University Press.

Escobar, Arturo. 1991. "Anthropology and the Development Encounter: The Making and Marketing of Development Anthropology." *American Ethnologist* 18 (4): 658–82.

Escobar, Arturo. 1994. *Encountering Development: The Making and Unmaking of the Third World*. Princeton, NJ: Princeton University Press.

Evans-Pritchard, E. E. 1946. "Applied Anthropology." *Africa* 16 (1): 92–98.

Fabian, Johannes. 1983. *Time and the Other: How Anthropology Makes Its Object*. New York: Columbia University Press.

Faudree, Paja. 2013. *Singing for the Dead: The Politics of Indigenous Revival in Mexico*. Durham, NC: Duke University Press.

Faudree, Paja. 2015. "Why X Doesn't Always Mark the Spot: Contested Authenticity in Mexican Indigenous Language Politics." *Semiotica* 203: 179–201.

Favela, Mariana. 2015. "Redrawing Power: #YoSoy132 and Overflowing Insurgencies." *Social Justice* 42 (3–4): 222–36.

Ferguson, James. 1990. *The Anti-Politics Machine: "Development," Depoliticization and Bureaucratic Power in Lesotho.* Minneapolis: University of Minnesota Press.

Ferguson, James, and Akhil Gupta. 2002. "Spatializing States: Toward an Ethnography of Neoliberal Governmentality." *American Ethnologist* 29 (4): 981–1002.

Fernández-Vega, Carlos. 2004. "Por ridículo precio, Iberdrola explota energía eólica en México." *La Jornada*, October 26. http://www.jornada.unam.mx/2004/10/26 /028a1eco.php?origen=opinion.php&fly=1.

Ferry, Elizabeth Emma. 2005. *Not Ours Alone: Patrimony, Value and Collectivity in Contemporary Mexico.* New York: Columbia University Press.

Fisher, William F. 1997. "Doing Good? The Politics and Antipolitics of NGO Practices." *Annual Review of Anthropology* 26: 439–64.

Fiske, Shirley, J., Susan A. Crate, Carole L. Crumley, Kathleen A. Galvin, Heather Lazrus, George Luber, Lisa Lucero, Anthony Oliver-Smith, Ben Orlove, Sarah Strauss, and Richard R. Wilk. 2014. *Changing the Atmosphere: Anthropology and Climate Change.* Final Report of the AAA Global Climate Change Task Force. Arlington, VA: American Anthropological Association.

Flauger, Jürgen, and Franz Hubik. 2016. "Electricity Prices in Free Fall." Handelsblatt Global, March 23. https://global.handelsblatt.com/edition/395/ressort/companies -markets/article/electricity-prices-in-a-free-fall.

Foster, George M. 1969. *Applied Anthropology.* Boston: Little, Brown.

Foucault, Michel. 1979. *Discipline and Punish.* New York: Pantheon.

Foucault, Michel. 1980. "The Confession of the Flesh." In *Power/Knowledge*, edited by Colin Gordon, 194–228. New York: Pantheon.

Foucault, Michel. 1984. *The History of Sexuality*, vol. 1, *Introduction.* New York: Vintage.

Foucault, Michel. 1993. "About the Beginnings of the Hermenuetics of the Self: Two Lectures at Dartmouth." *Political Theory* 21 (2): 198–227.

Foucault, Michel. 2000. "Governmentality." In *Essential Works of Foucault, 1954–1984*, vol. 3, *Power*, edited by James D. Faubion. New York: Free Press.

Foucault, Michel. 2004. *The Birth of Biopolitics.* New York: Picador.

Foweraker, Joe. 1990. "Popular Movements and Political Change in Mexico." In *Popular Movements and Political Change in Mexico*, edited by Joe Foweraker and Ann L. Craig, 3–22. Boulder, CO: Lynne Rienner.

Frank, Andre Gunder. 1969. *Latin America: Underdevelopment and Revolution.* New York: Monthly Review.

Franklin, Sarah, and Celia Roberts. 2006. *Born and Made: An Ethnography of Preimplantation Genetic Diagnosis.* Princeton, NJ: Princeton University Press.

Friedrich, Paul. 1965. "A Mexican Cacicazgo." *Ethnology* 4 (2): 190–209.

Friedrich, Paul. 1970. *Agrarian Revolt in a Mexican Village.* Englewood Cliffs, NJ: Prentice-Hall.

Fullwiley, Duana. 2006. "Biosocial Suffering: Order and Illness in Urban West Africa." *BioSocieties* 1 (4): 421–38.

Galbraith, Kate, and Asher Price. 2013. *The Great Texas Wind Rush: How George Bush, Ann Richards and a Bunch of Tinkerers Helped the Oil and Gas State Win the Race to Wind Power*. Austin: University of Texas Press.

Gledhill, John. 1995a. "The End of All Illusions? Neoliberalism, Transnational Economic Relations and Agrarian Reform in the Ciénega de Chapala, Michoacán." Unpublished manuscript. http://jg.socialsciences.manchester.ac.uk/cienega .pdf.

Gledhill, John. 1995b. *Neoliberalism, Transnationalization and Rural Poverty: A Case Study of Michoacán, Mexico*. Boulder, CO: Westview.

Gledhill, John. 2007. "Neoliberalism." In *A Companion to the Anthropology of Politics*, edited by David Nugent and Joan Vincent, 332–48. London: Blackwell.

Goertzen, Chris. 2010. *Made in Mexico: Tradition, Tourism and Political Ferment in Oaxaca*. Jackson: University of Mississippi Press.

Goldstein, Daniel M. 2004. *The Spectacular City: Violence and Performance in Urban Bolivia*. Durham, NC: Duke University Press.

Goldstein, Daniel M. 2005. "Flexible Justice: Neoliberal Violence and Self-Help Security in Bolivia." *Critique of Anthropology* 25: 389–411.

Goldwyn, David L., Neil R. Brown, and Megan Reilly Cayten. 2014. *Mexico's Energy Reform: Ready to Launch*. Washington, DC: Arsht Latin America Center of the Atlantic Council.

Gordillo, Gaston R. 2014. *Rubble: The Afterlife of Destruction*. Durham, NC: Duke University Press.

Graeber, David. 2002. "The New Anarchists." *New Left Review* 18: 61–73.

Graeber, David. 2010. "Against Kamikaze Capitalism: Oil, Climate Change and the French Refinery Blockades." *Shift*, November.

Graeber, David. 2011. *Debt: The First 5,000 Years*. Brooklyn, NY: Melville.

Grayson, George W. 2011. *Mexico: Narco-Violence and a Failed State?* New Brunswick, NJ: Transaction.

Greenberg, James B. 1997. "Caciques, Patronage, Factionalism and Variations among Local Forms of Capitalism." In *Citizens of the Pyramid: Essays on Mexican Political Culture*, edited by Wil G. Pansters, 309–36. Amsterdam: CEDLA.

Greenhalgh, Susan, and Edwin A. Winckler. 2005. *Governing China's Population: From Leninist to Neoliberal Biopolitics*. Stanford, CA: Stanford University Press.

Grewal, Inderpal, and Victoria Bernal, eds. 2014. *Theorizing NGOs: States, Feminisms and Neoliberalism*. Durham, NC: Duke University Press.

Gross, Toomas. 2012a. "Changing Faith: The Social Costs of Protestant Conversion in Rural Oaxaca." *Ethnos* 77 (3): 344–71.

Gross, Toomas. 2012b. "Incompatible Worlds? Protestantism and Indigenous Identity in Oaxaca." *Folklore* 51: 191–218.

Grupo Yansa. 2012. "Propuesta de Desarrollo Social de Ixtepec a través de un Proyecto Eólico Comunitario." August. Unpublished proposal.

Guerra, François-Xavier. 1992. "Los orígenes socio-culturales del caciquismo." *Anuario del IEHS* 7 (7): 181–83.

Guerrero, Gubidxa. "Opinión: La renuncia del Secretario de Gobierno." *Enfoque Diario*, April 15. Reproduced at Papeles del Sol. http://papelesdelsol.blogspot.com/2013/04/opinion-la-renuncia-del-secretario-de.html.

Gupta, Akhil. 2013. "Ruins of the Future." Paper presented at the American Anthropological Association Annual Meeting, Chicago.

Gupta, Akhil. 2015. "An Anthropology of Electricity from the Global South." *Cultural Anthropology* 30 (4): 555–68.

Gusterson, Hugh. 1996. *Nuclear Rites: A Weapons Laboratory at the End of the Cold War*. Berkeley: University of California Press.

Gutiérrez Brockington, Lolita. 1989. *The Leverage of Labor: Managing the Cortés Haciendas in Tehuantepec, 1588–1688*. Durham, NC: Duke University Press.

Habermas, Jürgen. 1989. *The Structural Transformation of the Public Sphere*. Cambridge, MA: MIT Press.

Haenn, Nora. 2006. "The Changing and Enduring Ejido: A State and Regional Examination of Mexico's Land Tenure Counter-Reforms." *Land Use Policy* 23: 136–46.

Hale, Charles. 1991. *La transformación del liberalismo mexicano en el último cuarto del siglo XIX*. Mexico City: Editorial Vuelta.

Hale, Charles R. 2006. *Más que un Indio (More Than an Indian): Racial Ambivalence and Neoliberal Multiculturalism in Guatemala*. Santa Fe, NM: School of American Research Press.

Hanson, Paul W. 2007. "Governmentality, Language Ideology and the Production of Needs in Malagasy Conservation and Development." *Cultural Anthropology* 22 (2): 244–84.

Haraway, Donna. 1985. "A Cyborg Manifesto: Science, Technology, and Socialist-Feminism in the Late Twentieth Century." *Socialist Review* 80: 65–108.

Haraway, Donna. 2003. *The Companion Species Manifesto: Dogs, People and Significant Otherness*. Chicago: Prickly Paradigm.

Hardt, Michael, and Antonio Negri. 2000. *Empire*. Cambridge, MA: Harvard University Press.

Harman, Graham. 2002. *Tool-Being: Heidegger and the Metaphysics of Objects*. Chicago: Open Court.

Hartigan, John. 2013. "Mexican Genomics and the Roots of Racial Thinking." *Cultural Anthropology* 28 (3): 372–95.

Harvey, David. 2005. *A Brief History of Neoliberalism*. Oxford: Oxford University Press.

Harvey, David. 2007. "Neoliberalism as Creative Destruction." *Annals of the American Academy of Political and Social Science* 610 (March): 22–44.

Harvey, Penny, Caspar Bruun Jensen, and Atsuro Morita, eds. 2017. *Infrastructures and Social Complexity*. London: Routledge.

Harvey, Penny, and Hannah Knox. 2015. *Roads: An Anthropology of Infrastructure*. Ithaca, NY: Cornell University Press.

Haugerud, Angelique. 2013. *No Billionaire Left Behind: Satirical Activism in America*. Stanford, CA: Stanford University Press.

Hausman, William J., and John L. Neufeld. 1997. "The Rise and Fall of the American & Foreign Power Company: A Lesson from the Past?" *Electricity Journal* 10 (1): 46–53.

Hellman, Judith Adler. 1997. "Structural Adjustment in Mexico and the Dog That Didn't Bark." CERLAC Working Paper Series, York University. April. http://www.yorku.ca/cerlac/documents/Hellman.pdf.

Henning, Annette. 2005. "Climate Change and Energy Use." *Anthropology Today* 21 (3): 8–12.

Hermesdorff, M. G. 1862. "On the Isthmus of Tehuantepec." *Journal of the Royal Geographical Society* 32: 536–54.

Hoben, Allan. 1982. "Anthropologists and Development." *Annual Review of Anthropology* 11: 349–75.

Hoffmann, Julia. 2012. "The Social Power of Wind: The Role of Participation and Social Entrepreneurship in Overcoming Barriers for Community Wind Farm Development, Lessons from the Ixtepec Community Wind Farm Project in Mexico." MA thesis, Lund University.

Holmes, Douglas R. 2000. *Integral Europe: Fast Capitalism, Multiculturalism, Neofascism.* Princeton, NJ: Princeton University Press.

Howe, Cymene, and Dominic Boyer. 2015. "Aeolian Politics." *Distinktion* 16 (1): 31–48.

Howe, Cymene, and Dominic Boyer. 2016. "Aeolian Extractivism and Community Wind in Southern Mexico." *Public Culture* 28 (2): 215–35.

Howe, Cymene, Jessica Lockrem, Hannah Appel, Edward Hackett, Dominic Boyer, Randal Hall, Matthew Schneider-Mayerson, et al. 2015. "Paradoxical Infrastructures: Ruins, Retrofit, and Risk." *Science, Technology and Human Values* 41 (3): 547–65.

Howell, Jayne. 2009. "Vocation or Vacation? Perspectives on Teachers' Union Struggles in Southern Mexico." *Anthropology of Work Review* 30 (3): 87–98.

Hu-DeHart, Evelyn. 1988. "Peasant Rebellion in the Northwest: The Yaqui Indians of Sonora, 1740–1976." In *Riots, Rebellion and Revolution: Rural Social Conflict in Mexico*, edited by Friedrich Katz, 141–75. Princeton, NJ: Princeton University Press.

Hughes, Thomas. 1983. *Networks of Power: Electrification in Western Society, 1880–1930.* Baltimore: Johns Hopkins University Press.

Hull, Matthew S. 2012. *Government of Paper: The Materiality of Bureaucracy in Urban Pakistan.* Berkeley: University of California Press.

Hunn, David. 2016. "Mexico Could Open Shale Fields to US Drillers Next Year." FuelFix, September 23. http://fuelfix.com/blog/2016/09/23/mexico-could-open-shale-fields-to-u-s-drillers-next-year/.

Hymes, Dell, ed. 1972. *Reinventing Anthropology.* New York: Pantheon.

Iliff, Laurence. 2015. "Mexico's Pemex Lands Pipeline Deal with BlackRock, First Reserve." *Wall Street Journal*, March 26. http://www.wsj.com/articles/mexicos-pemex-lands-pipeline-deal-with-blackrock-first-reserve-1427405627.

Jamieson, Dale. 2001. "Climate Change and Global Environmental Justice." In *Changing the Atmosphere: Expert Knowledge and Global Environmental Governance*, edited by Clark A. Miller and Paul N. Edwards, 287–307. Cambridge, MA: MIT Press.

Jamieson, Dale. 2011. "Energy, Ethics and the Transformation of Nature." *The Ethics of Global Climate Change*, edited by Denis G. Arnold, 16–37. Cambridge: Cambridge University Press.

Johnston, Barbara Rose, Susan E. Dawson, and Gary E. Madsen. 2010. "Uranium Mining and Milling: Navajo Experiences in the American Southwest." In *Indians and Energy*, edited by Sherry Smith and Brian Frehner, 97–116. Santa Fe, NM: SAR Press.

Jorgensen, Joseph G., ed. 1984. *Native Americans and Energy Development II*. Boston: Anthropology Resource Center.

Jorgensen, Joseph G. 1990. *Oil Age Eskimos*. Berkeley: University of California Press.

Jorgensen, Joseph G., Richard O. Clemmer, Ronald L. Little, Nancy J. Owens, and Lynn A. Robbins. 1978. *Native Americans and Energy Development*. Cambridge, MA: Anthropology Resource Center.

Joseph, Gilbert M. 1988. *Revolution from Without: Yucatán Mexico and the United States, 1880–1924*. Durham, NC: Duke University Press.

Joyce, Stephanie. 2016. "Do We Need Coal to Keep the Lights On?" Audio. Wyoming Public Radio, June 10. Inside Energy. http://insideenergy.org/2016/06/10/do-we-need-coal-to-keep-the-lights-on/.

Jusionyte, Ieva. 2015. "States of Camouflage." *Cultural Anthropology* 30 (1): 113–38.

Kallis, Giorgos. 2011. "In Defence of Degrowth." *Ecological Economics* 70: 873–80.

Kallis, Giorgos. 2018. *Degrowth*. New York: Columbia University Press.

Karim, Lamia. 2001. "Politics of the Poor? NGOs and Grass-Roots Political Mobilization in Bangladesh." *PoLAR* 24 (1): 92–107.

Katz, Friedrich. 1988. "Rural Rebellions after 1810." In *Riots, Rebellion and Revolution: Rural Social Conflict in Mexico*, edited by Friedrich Katz, 521–60. Princeton, NJ: Princeton University Press.

Katz, Friedrich. 2004. *De Díaz a Madero*. Mexico City: Ediciones Era.

Kauanui, J. Kēhaulani. 2008. *Hawaiian Blood: Colonialism and the Politics of Sovereignty and Indigeneity*. Durham, NC: Duke University Press.

Kaufman, Purcell S. 1981. "Mexico: Clientelism, Corporatism and Political Stability." In *Political Clientelism, Patronage and Development*, edited by Shmuel N. Eisenstadt and Rene Lemarchand. London: Sage.

Kearney, Michael. 1986. "From the Invisible Hand to the Visible Feet: Anthropological Studies of Migration and Development." *Annual Review of Anthropology* 15: 331–61.

Kearney, Michael. 2004. *Changing Fields of Anthropology: From Local to Global*. New York: Rowman & Littlefield.

Kelly, John H., et al. 2010. "Indigenous Territoriality at the End of the Social Property Era in Mexico." *Journal of Latin American Geography* 9 (3): 161–81.

Kelly, Patty. 2008. *Lydia's Open Door: Inside Mexico's Most Modern Brothel*. Berkeley: University of California Press.

Kemper, Robert V. 1977. *Migration and Adaptation: Tzintzuntzan Peasants in Mexico City*. Beverly Hills, CA: Sage.

Kerevel, Yann P. 2010. "The Legislative Consequences of Mexico's Mixed-Member Electoral System, 2000–2009." *Electoral Studies* 29 (4): 691–703.

King, Stacie M. 2012. "Hidden Transcripts, Contested Landscapes and Long-Term Indigenous History in Oaxaca, Mexico." In *Decolonizing Indigenous Histories: Exploring Prehistoric/Colonial Transitions in Archaeology*, edited by Maxine Oland, Siobhan M. Hart, and Liam Frink, 230–66. Tucson: University of Arizona Press.

Kirksey, S. Eben, and Stefan Helmreich. 2010. "The Emergence of Multispecies Ethnography." *Cultural Anthropology* 25 (4): 545–76.

Klein, Naomi. 2015. *This Changes Everything: Capitalism vs. the Climate*. New York: Simon & Schuster.

Klieman, Kairn. 2008. "Oil, Politics, and Development in the Formation of a State: The Congolese Petroleum Wars, 1963–68. *International Journal of African Historical Studies* 41 (2): 169–202.

Klumbyte, Neringa. 2011. "Political Intimacy: Power, Laughter, and Coexistence in Late Soviet Lithuania." *East European Politics and Societies* 25 (4): 658–77.

Knight, Alan. 2002. *Mexico: The Colonial Era*. Vol. 2. Cambridge: Cambridge University Press.

Knight, Alan, and Wil G. Pansters, eds. 2005. *Caciquismo in Twentieth-Century Mexico*. London: Institute for the Study of the Americas, University of London.

Knight, Daniel M. 2015. "Wit and Greece's Economic Crisis: Ironic Slogans, Food, and Antiausterity Sentiments." *American Ethnologist* 42 (2): 230–46.

Komives, Kristin, Todd M. Johnson, Jonathan D. Halpern, José Luis Aburto, and John R. Scott. 2009. *Residential Electricity Subsidies in Mexico: Exploring Options for Reform and for Enhancing the Impact on the Poor*. World Bank Working Paper 160. Washington, DC: World Bank.

Kourí, Emilio H. 2002. "Interpreting the Expropriation of Indian Pueblo Lands in Porfirian Mexico: The Unexamined Legacies of Andrés Molina Enríquez." *Hispanic American Historical Review* 82 (1): 69–117.

Kraemer Bayer, Gabriela. 2008. *Autonomía de los zapotecos del Istmo: Relaciones de poder y cultura política*. Mexico City: CONACYT, Plaza y Valdés.

Kruijt, Dirk. 2012. *Drugs, Democracy and Security: The Impact of Organized Crime on the Political System of Latin America*. The Hague: NIMD.

Kruse, John, Judith Kleinfeld, and Robert Travis. 1982. "Energy Development on Alaska's North Slope: Effects on the Inupiat Population." *Human Organisation* 41 (2): 95, 97–106.

Kuriyan, Renee, and Isha Ray. 2009. "Outsourcing the State? Public-Private Partnerships and Information Technologies in India." *World Development* 37 (10): 1663–73.

Lacey, Marc. 2009. "Mexico Says It Is Closing a Provider of Electricity." *New York Times*, October 12. http://www.nytimes.com/2009/10/12/world/americas/12mexico .html.

Lakoff, Andrew, and Stephen J. Collier, eds. 2008. *Biosecurity Interventions: Global Health and Security in Question*. New York: Columbia University Press.

Land, Nick. 2011. *Fanged Noumena: Collected Writings 1987–2007*. Edited by Robin Mackay and Ray Brassier. Falmouth, UK: Urbanomic.

Larkin, Brian. 2013. "The Politics and Poetics of Infrastructure." *Annual Review of Anthropology* 42: 327–43.

Latour, Bruno. 1988. *The Pasteurization of France*. Translated by Alan Sheridan and John Law. Cambridge, MA: Harvard University Press.

Latour, Bruno. 2004. "Why Has Critique Run Out of Steam? From Matters of Fact to Matters of Concern." *Critical Inquiry* 30 (Winter): 225–48.

Lemke, Thomas. 2002. "Foucault, Governmentality and Critique." *Rethinking Marxism* 14 (3): 49–64.

Lévi-Strauss, Claude. 1966. *The Savage Mind*. Chicago: University of Chicago Press.

Li, Tania Murray. 2007. *The Will to Improve: Governmentality, Development, and the Practice of Politics*. Durham, NC: Duke University Press.

Liffman, Paul M. 2011. *Huichol Territory and the Mexican Nation: Indigenous Ritual, Land Conflict, and Sovereignty Claims*. Tucson: University of Arizona Press.

LiPuma, Edward, and Benjamin Lee. 2004. *Financial Derivatives and the Globalization of Risk*. Durham, NC: Duke University Press.

Lizama Quijano, Jesús. 2006. *La Guelaguetza en Oaxaca: Fiesta, relaciones interétnicas y procesos de construcción simbólica en el context urbano*. Mexico City: CIESAS.

Loaeza, Soledad. 2006. "Problems of Political Consolidation in Mexico." In *Changing Structure of Mexico: Political, Social and Economic Prospects*, edited by Laura Randall, 32–48. New York: Routledge.

Loewe, Ronald. 2011. *Maya or Mestizo: Nationalism, Modernity and Its Discontents*. Toronto: University of Toronto Press.

Lomnitz, Claudio. 2001. *Deep Mexico, Silent Mexico: An Anthropology of Nationalism*. Minneapolis: University of Minnesota Press.

Lomnitz, Claudio. 2005. "Sobre reciprocidad negativa." *Revista de Antropología Social* 14: 311–39.

Lomnitz, Claudio. 2008. "Narrating the Neoliberal Moment: History, Journalism, Historicity." *Public Culture* 20 (1): 39–56.

Lomnitz, Larissa. 1977. *Networks and Marginality: Life in a Mexican Shantytown*. New York: Academic.

López Villalobos, Manuel. 2004. *Juchitán Crisis de un Proyecto Político COCEI: Orígen, Consolidación y Crisis 1974–1998*. Mexico City: Universidad Autónoma Metropolitana.

Love, Thomas. 2008. "Anthropology and the Fossil Fuel Era." *Anthropology Today* 24 (2): 3–4.

Love, Thomas, and Anna Garwood. 2011. "Wind, Sun and Water: Complexities of Alternative Energy Development in Rural Northern Peru." *Rural Society* 20: 294–307.

Luke, Timothy W. 1999. "Environmentality as Green Governmentality." In *Discourses of the Environment*, edited by Eric Darier, 121–51. Malden, MA: Blackwell.

Lynch, Barbara. 1982. *The Vicos Experiment: A Study of the Impacts of the Cornell-Peru Project in a Highland Community*. Washington, DC: USAID.

Mackay, Robin, ed. 2014. *#ACCELERATE: The Accelerationist Reader*. Falmouth, UK: Urbanomic.

MacKenzie, Donald. 2008. *An Engine, Not a Camera: How Financial Models Shape Markets*. Cambridge, MA: MIT Press.

MacKenzie, Donald, Febian Muniesa, and Lucia Siu. 2007. *Do Economists Make Markets? On the Performativity of Economics*. Princeton, NJ: Princeton University Press.

Malette, Sebastien. 2009. "Foucault for the Next Century: Eco-Governmentality." In *A Foucault for the 21st Century: Governmentality, Biopolitics and Discipline in the New Millennium*, edited by Sam Binkley and Jorge Capetillo, 221–39. Cambridge: Cambridge Scholars.

Malinowski, Bronislaw. 1929. "Practical Anthropology." *Africa* 2 (1): 22–38.

Malm, Andreas. 2013. "The Origins of Fossil Capital: From Water to Steam in the British Cotton Industry." *Historical Materialism* 21 (1): 15–68.

Malm, Andreas, and Alf Hornborg. 2014. "The Geology of Mankind? A Critique of the Anthropocene Narrative." *Anthropocene Review* 1 (1): 62–69.

Manzagol, Nilay, and Tyler Hodge. 2016. "Mexico Electricity Market Reforms Attempt to Reduce Costs and Develop New Capacity." US Energy Information Administration. July 5. http://www.eia.gov/todayinenergy/detail.cfm?id=26932.

Martínez López, Aurelio. 1966. *Historia de la Intervencion Francesa en el Estado de Oaxaca*. Mexico City.

Martínez Vásquez, Victor Raúl. 2007. *Autoritarismo, Movimiento Popular y Crisis Política: Oaxaca 2006*. Oaxaca: Universidad Autónoma Benito Juárez de Oaxaca, CAMPO, EDUCA, CDPE.

Martínez Vásquez, Victor Raúl. 2008. "Crisis política y repression en Oaxaca." *El Cotidiano* 148: 45–61.

Marx, Karl. (1861) 1974. *Grundrisse der Kritik der Politischen Ökonomie*. Berlin: Dietz Verlag.

Masco, Joseph. 2006. *Nuclear Borderlands: The Manhattan Project in Post–Cold War New Mexico*. Princeton, NJ: Princeton University Press.

Mason, Arthur. 2007. "The Rise of Consultant Forecasting in Liberalized Natural Gas Markets." *Public Culture* 19 (2): 367–79.

Mason, Arthur, and Maria Stoilkova. 2012. "Corporeality of Consultant Expertise in Arctic Natural Gas Development." *Journal of Northern Studies* 6 (2): 83–96.

Massumi, Brian. 2015. *Ontopower: War, Powers, and the State of Perception*. Durham, NC: Duke University Press.

Mathews, Andrew S. 2011. *Instituting Nature: Authority, Expertise, and Power in Mexican Forests*. Cambridge, MA: MIT Press.

Matsutake Worlds Research Group. 2009. "A New Form of Collaboration in Cultural Anthropology: Matsutake Worlds." *American Ethnologist* 36 (2): 380–403.

May, Robert E. 1973. *The Southern Dream of Caribbean Empire, 1854–1861*. Baton Rouge: Louisiana State University Press.

McDonald, James H. 1999. "The Neoliberal Project and Governmentality in Rural Mexico: Emergent Farmer Organization in the Michoacán Highlands." *Human Organization* 58 (3): 274–84.

McNeish, John-Andrew, and Owen Logan, eds. 2012. *Flammable Societies: Studies on the Socio-Economics of Oil and Gas*. London: Pluto.

Mead, Margaret, ed. 1953. *Cultural Patterns and Technical Change*. Paris: UNESCO.

Mecham, J. Lloyd. 1938. "The Origins of Mexican Federalism." *Hispanic American Historical Review* 18 (2): 164–82.

Meillasoux, Quentin. 2008. *After Finitude: An Essay on the Necessity of Contingency.* Translated by Ray Brassier. London: Continuum.

Melgar, Lourdes, Alejandro Díaz de Leon, and Victor Luque. 2015. "Mexico's New Energy Industry: Investing in the Transformation." Mexico City: SENER.

Méndez, Enrique, and Roberto Garduño. 2012. "Proponen plan eólico alterno para Oaxaca." *La Jornada*, October 19, 37.

Merchant, David, and Paul Rich. 2003. "Prospects for Mexican Federalism: Roots of the Policy Issues." *Policy Studies Journal* 31 (4): 661–67.

Merlan, Francesca. 2009. "Indigeneity: Global and Local." *Current Anthropology* 50 (3): 303–33.

Meyer, Niels I. 2004. "Development of Danish Wind Power Market." *Energy and Environment* 15 (4): 657–73.

Mimiaga Sosa, Fernando. 2009. "Corredor Eólico del Istmo de Tehuantepec." Power-Point presentation. Secretaría de Economía del Gobierno del Estado de Oaxaca.

Mitchell, Timothy. 2009. "Carbon Democracy." *Economy and Society* 38 (3): 399–432.

Mitchell, Timothy. 2011. *Carbon Democracy: Political Power in the Age of Oil.* New York: Verso.

Molé, Noelle J. 2013. "Trusted Puppets, Tarnished Politicians: Humor and Cynicism in Berlusconi's Italy." *American Ethnologist* 40 (2): 288–99.

Moore, Jason W. 2015. *Capitalism in the Web of Life.* London: Verso.

Moreton-Robinson, Aileen. 2015. *The White Possessive: Property, Power, and Indigenous Sovereignty.* Minneapolis: University of Minnesota Press.

Morton, Timothy. 2013. *Hyperobjects: Philosophy and Ecology after the End of the World.* Minneapolis: University of Minnesota Press.

Mouffe, Chantal. 2005. *On the Political.* London: Routledge.

Muehlebach, Andrea. "Time of Monsters." Theorizing the Contemporary, *Cultural Anthropology* website. October 27, 2016. https://culanth.org/fieldsights/979-time-of -monsters.

Muehlmann, Shaylih. 2013. *Where the River Ends: Contested Indigeneity in the Mexican Colorado Delta.* Durham, NC: Duke University Press.

Mueller, Adele. 1986. "The Bureaucratization of Development Knowledge: The Case of Women in Development." *Resources for Feminist Research* 15 (1): 3–6.

Muir, Sarah. 2015. "The Currency of Failure: Money and Middle-Class Critique in Post-Crisis Buenos Aires." *Cultural Anthropology* 30 (2): 310–35.

Münch Galindo, Guido. 2006. *La organización ceremonial de Tehuantepec y Juchitán.* Mexico City: UNAM—Instituto de Investigaciones Antropológicas.

Nader, Laura. 1980. *Energy Choices in a Democratic Society: A Resource Group Study for the Synthesis Panel of the Committee on Nuclear Alternative Energy Systems for the US National Academy of Sciences.* Washington, DC: National Academy of Sciences.

Nader, Laura. 1981. "Barriers to Thinking New about Energy." *Physics Today* 34 (9): 99–104.

Nader, Laura. 2004. "The Harder Path—Shifting Gears." *Anthropological Quarterly* 77 (4): 771–91.

Nader, Laura, ed. 2010. *The Energy Reader.* Oxford: Wiley-Blackwell.

Nader, Laura, and Stephen Beckerman. 1978. "Energy as It Relates to the Quality and Style of Life." *Annual Review of Energy* 3: 1–28.

Nahmad, Salomón. 2007. "Situación social y política de México y de Oaxaca al final del gobierno de Vicente Fox y principios del gobierno de Felipe Calderón." LASA *Forum* 38 (2): 24–26.

Nahmad Sittón, Salomón. 2011. "Informe final para el consejo oaxaqueño de ciencia y tecnología (COCyT) del CONACyT, proyecto 123396, el impacto social del uso del recurso eólico." Oaxaca: CIESAS Unidad Pacifico Sur.

Nash, June. 1981. "Ethnographic Aspects of the World Capitalist System." *Annual Review of Anthropology* 10: 393–423.

Nava Morales, Elena. 2015. "Radio Totopo y sus jóvenes: Instituciones comunitarias y procesos de Resistencia." *Antipoda Revista de Antropología y Arqueología* 23: 89–113. http://dx.doi.org/10.7440/antipoda23.2015.05.

Negarestani, Reza. 2008. *Cyclonopedia: Complicity with Anonymous Materials.* Melbourne: Re.Press.

NERC. 2014. *Essential Reliability Services: A Tutorial for Maintaining Bulk Power System Reliability and Adapting to a Changing Resource Mix.* Atlanta: North American Electric Reliability Corporation.

Niblo, Stephen R. 1999. *Mexico in the 1940s: Modernity, Politics, and Corruption.* New York: Rowman & Littlefield.

Nordstrom, Jean Maxwell, et al. 1977. *The Northern Cheyenne Tribe and Energy Developments in Southeastern Montana*, vol. 1, *Social and Cultural Investigations.* Lame Deer, MT: Northern Cheyenne Research Project.

Norget, Kristin. 2006. *Days of Death, Days of Life: Ritual in the Popular Culture of Oaxaca.* New York: Columbia University Press.

Norget, Kristin. 2010. "A Cacophony of Autochthony: Representing Indigeneity in Oaxacan Popular Mobilization." *Journal of Latin American and Caribbean Anthropology* 15: 116–43.

Nugent, Daniel, and Ana María Alonso. 1994. "Multiple Selective Traditions in Agrarian Reform and Agrarian Struggle: Popular Culture and State Formation in the Ejido of Namiquipa, Chihuahua." In *Everyday Forms of State Formation: Revolution and the Negotiation of Rule in Modern Mexico*, edited by Gilbert M. Joseph and Daniel Nugent, 209–46. Durham, NC: Duke University Press.

Nye, David E. 1992 *Electrifying America: Social Meanings of a New Technology.* Cambridge, MA: MIT Press.

Nye, David E. 2010. *When the Lights Went Out: A History of Blackouts in America.* Cambridge, MA: MIT Press.

Oceransky, Sergio. 2010. "Fighting the Enclosure of Wind: Indigenous Resistance to the Privatization of the Wind Resource in Southern Mexico," In *Sparking a Worldwide Energy Revolution: Social Struggles in the Transition to a Post-Petrol World*, edited by Kolya Abramsky, 505–22. Oakland, CA: AK.

Ochoa, Enrique C. 2001. "Neoliberalism, Disorder, and Militarization in Mexico." *Latin American Perspectives* 28 (4): 148–59.

O'Neill, John J. 1940. "Enter Atomic Power." *Harper's* 181 (June): 1–10.

Ong, Aihwa, and Stephen J. Collier, eds. 2005. *Global Assemblages: Technology, Politics and Ethics as Anthropological Problems.* New York: Wiley-Blackwell.

Oreskes, Naomi, and Erik M. Conway. 2013. "The Collapse of Western Civilization: A View from the Future." *Daedalus* 142 (1): 40–58.

Ortiz Rubén, Anzaldo, Avendaño Arenas Itzel, Bonilla Mendoza Javier, Mejía Uribe Iván, and Tierranueva León Erandi. 2014. *La Resistencia de la Asamblea Popular del Pueblo Juchiteco contra la Construcción del Megaproyecto Eólico: Identidad y Repertorio de Acción.* Mexico City: Universidad Autónoma Metropolitana.

Oudijk, Michel R. 2000. *Historiography of the Bènizàa: e Postclassic and Early Colonial Periods (1000–1600 AD).* CNWS Publications 84. Leiden: Research School of Asian, African, and Amerindian Studies, University of Leiden.

Oudijk, Michel. 2008. "Una nueva historia zapoteca." In *Pictografía y escritura alfabética en Oaxaca*, edited by Sebastián van Doesburg, 89–116. Oaxaca: Instituto Estatal de Educación Pública de Oaxaca.

Özden-Schilling, Canay. 2015. "Economy Electric." *Cultural Anthropology* 30 (4): 578–88.

Paley, Julia. 2001. "The Paradox of Participation: Civil Society and Democracy in Chile." *PoLAR* 24 (1): 1–12.

Pansters, Wil G. 2005. "Goodbye to the Caciques? Definition, the State and the Dynamics of Caciquismo in Twentieth-Century Mexico." In *Caciquismo in Twentieth-Century Mexico*, edited by Alan Knight and Wil G. Pansters, 349–76. London: Institute for the Study of the Americas, University of London.

Pansters, Wil G., ed. 2012. *Violence, Coercion and State-Making in Twentieth-Century Mexico: The Other Half of the Centaur.* Stanford, CA: Stanford University Press.

Patel, Raj, and Jason W. Moore. 2017. *A History of the World in Seven Cheap Things: A Guide to Capitalism, Nature and the Future of the Planet.* Berkeley: University of California Press.

Perramond, Eric. 2008. "The Rise, Fall and Reconfiguration of the Mexican *Ejido*." *Geographical Review* 98 (3): 356–71.

Petryna, Adriana. 2002. *Life Exposed: Biological Citizens after Chernobyl.* Princeton, NJ: Princeton University Press.

Postero, Nancy Grey. 2006. *Now We Are Citizens: Indigenous Politics in Postmulticultural Bolivia.* Palo Alto, CA: Stanford University Press.

Potter, Robert D. 1940. "Is Atomic Power at Hand?" *Scientific Monthly* 50 (6): 571–74.

Povinelli, Elizabeth. 2002. *The Cunning of Recognition: Indigenous Alterities and the Making of Australian Multiculturalism.* Durham, NC: Duke University Press.

Povinelli, Elizabeth. 2011. *Economies of Abandonment: Social Belonging and Endurance in Late Liberalism.* Durham, NC: Duke University Press.

Povinelli, Elizabeth. 2016. *Geontologies: A Requiem to Late Liberalism.* Durham, NC: Duke University Press.

Powell, Dana E., and Dáilan J. Long. 2010. "Landscapes of Power: Renewable Energy Activism in Diné Bikéyah." In *Indians and Energy: Exploitation and Opportunity in the American Southwest*, edited by Sherry L. Smith and Brian Frehner, 231–62. Santa Fe, NM: SAR Press.

Price, David H. 2016. *Cold War Anthropology: The CIA, the Pentagon, and the Growth of Dual Use Anthropology*. Durham, NC: Duke University Press.

Purnell, Jennie. 2005. "The Chegomista Rebellion in Juchitán, 1911–1912: Rethinking the Role of 'Traditional' Caciques in Resisting Central State Power." In *Caciquismo in Twentieth Century Mexico*, edited by Alan Knight and Wil G. Pansters, 51–70. London: Institute for the Study of the Americas, University of London.

Raat, William. 1975. *El positivismo durante el porfiriato, 1876–1910*. Mexico City: SEP.

Rabinow, Paul, and Nikolas Rose. 2006. "Biopower Today." *BioSocieties* 1 (2): 195–217.

Rancière, Jacques. 1998. *Disagreement*. Minneapolis: University of Minnesota Press.

Rancière, Jacques. 2001. "Ten Theses on Politics." *Theory and Event* 5 (3): 27–44.

Rappaport, Roy 1975. "The Flow of Energy in Agricultural Society." In *Biological Anthropology: Readings from Scientific American*, edited by Solomon Katz, 371–87. San Francisco: Freeman.

Redfield, Peter. 2005. "Doctors, Borders and Life in Crisis." *Cultural Anthropology* 20 (3): 328–61.

Reyna, Stephen, and Andrea Behrends. 2008. "The Crazy Curse and Crude Domination: Toward an Anthropology of Oil." *Focaal* 52 (Winter): 3–17.

Richard, Analiese M. 2009. "Mediating Dilemmas: Local NGOs and Rural Development in Neoliberal Mexico." *PoLAR* 32 (2): 166–94.

Rippy, J. Fred. 1920. "Diplomacy of the United States and Mexico Regarding the Isthmus of Tehuantepec, 1848–1860." *Mississippi Valley Historical Review* 6 (4): 503–31.

Ristow, Colby Nolan. 2008. "From Repression to Incorporation in Revolutionary Mexico: Identity Politics, Cultural Mediation, and Popular Revolution in Oaxaca, 1910–1920." PhD diss., University of Chicago.

Robbins, Lynn. 1980. *The Socioeconomic Impacts of the Proposed Skagit Nuclear Power Plant on the Skagit System Cooperative Tribes*. Bellingham, WA: Lord.

Robbins, Lynn. 1984. "Energy Developments and the Navajo Nation: An Update." In *Native Americans and Energy Development*, edited by Joseph Jorgensen. Boston: Anthropology Resource Center.

Rochlin, James F. 1997. *Redefining Mexican "Security": Society, State, and Region under NAFTA*. Boulder, CO: Lynne Rienner.

Rodríguez, Nemesio J. 2002. "Pueblos indios, globalización y desarrollo (ensayo para entender)." In *Estado del desarrollo económico y social de los pueblos indígenas de México. Segundo informe*. Mexico City: Instituto Nacional Indigenista-PNUD.

Rogers, Douglas. 2015. *The Depths of Russia: Oil, Power, and Culture after Socialism*. Ithaca, NY: Cornell University Press.

Rojinsky, David. 2008. "Manso de Contreras' *Relación* of the Tehuantepec Rebellion (1660–1661): Violence, Counter-Insurgency Prose and the Frontiers of Colonial Justice." In *Border Interrogations: Questioning Spanish Frontiers*, edited by Benita Sampedro Vizcaya and Simon Doubleday, 198–203. New York: Berghahn.

Rose, Nikolas, Pat O'Malley, and Mariana Valverde. 2006. "Governmentality." *Annual Review of Law and Social Science* 2: 83–104.

Roseberry, William. 1988. "Political Economy." *Annual Review of Anthropology* 17: 161–85.

Royce, Anna Peterson. 1991. "Music, Dance, and Fiesta: Definitions of Isthmus Zapotec Community." *Latin American Anthropology Review* 3: 51–60.

Rubin, Jeffrey W. 1997. *Decentering the Regime: Ethnicity, Radicalism and Democracy in Juchitán, Mexico*. Durham, NC: Duke University Press.

Rudnyckyj, Daromir. 2010. *Spiritual Economies: Islam, Globalization and the Afterlife of Development*. Ithaca, NY: Cornell University Press.

Ruiz Medrano, Carlos Rubén. 2005. "La Resistencia Indígena en la Sierra de Tututepeque, Nueva España, Durante la Segunda Mitad del Siglo XVIII." *Mesoamérica* 47 (January–December): 23–46.

Sader, Emir. 2008. "The Weakest Link? Neoliberalism in Latin America." *New Left Review* 52 (July/August): 5–31.

Salinas Sandoval, Maria del Carmen. 2014. *El Primer Federalismo in el Estado de México 1824–1835*. Toluca, México: El Colegio Mexiquense.

Sánchez Prado, Ignacio M. 2015. "Democracy, Rule of Law, a 'Loving Republic,' and the Impossibility of the Political in Mexico." Translated by Ariel Wind. *Política Común* 7. http://quod.lib.umich.edu/p/pc/12322227.0007.004?view=text;rgn=main.

Sánchez Santiró, Ernest. 2001. *Azúcar y poder: Estructura socioeconómica de las alcaldías mayores de Cuernavaca y Cuautla de Amilpas, 1730–1821*. Mexico City: Editorial Praxis, Universidad Autónoma del Estado de Morelos.

Sawyer, Suzana. 2004. *Crude Chronicles: Indigenous Politics, Multinational Oil, and Neoliberalism in Ecuador*. Durham, NC: Duke University Press.

Sawyer, Suzana. 2007. "Empire/Multitude—State/Civil Society: Rethinking Topographies of Power through Transnational Connectivity in Ecuador and Beyond." *Social Analysis* 51 (2): 64–85.

Sawyer, Suzana, and Terence Gomez, eds. 2012. *The Politics of Resource Extraction: Indigenous Peoples, Corporations and the State*. London: Palgrave Macmillan.

Saynes-Vásquez, Floria E. 2002. "Zapotec Language Shift and Reversal in Juchitán, México." PhD diss., University of Arizona.

Scheer, Hermann. 2002. *The Solar Economy*. New York: Earthscan.

Scheer, Hermann. 2006. *Energy Autonomy: The Economic, Social and Technological Case for Renewable Energy*. New York: Earthscan.

Schiller, Naomi. 2011. "Catia Sees You: Community Television, Clientelism, and Participatory Statemaking in the Chávez Era." In *Venezuela's Bolivarian Democracy: Participation, Politics and Culture under Chávez*, edited by David Smilde and Daniel Hellinger, 104–30. Durham, NC: Duke University Press.

Schuller, Mark. 2009. "Gluing Globalization: NGOs as Intermediaries in Haiti." *PoLAR* 32 (1): 84–104.

Schwegler, Tara A. 2008. "Take It from the Top (Down)? Rethinking Neoliberalism and Political Hierarchy in Mexico." *American Ethnologist* 35 (4): 682–700.

Secretaría de Energía. 2015. "Desarrollo sostenible." Gobierno de Mexico. July 16. http://www.gob.mx/sener/articulos/desarrollo-sostenible.

Selee, Andrew. 2011. *Decentralization, Democratization, and Informal Power in Mexico*. University Park: Pennsylvania State University Press.

Shah, Jigar (@JigarShahDC). 2015. "There is no such thing as 'baseload,' #nuclear/#coal runs too much, #natgas too little, they mean 'dispatchable' @benpaulos @atrembath." Twitter, August 15, 2:02 PM. https://twitter.com/jigarshahdc/status/632658752987111424.

Sharma, Aradhana. 2006. "Crossbreeding Institutions, Breeding Struggle: Women's Empowerment, Neoliberal Governmentality and State (Re)Formation in India." *Cultural Anthropology* 21 (1): 60–95.

Sharma, Aradhana, and Akhil Gupta, eds. 2006. *The Anthropology of the State: A Reader*. Malden, MA: Blackwell.

Shore, Cris, and Susan Wright, eds. 2003. *Anthropology of Policy: Critical Perspectives on Governance and Power*. London: Routledge.

Simpson, Audra. 2014. *Mohawk Interruptus: Political Life across the Borders of Settler States*. Durham, NC: Duke University Press.

Simpson, Audra. 2016. "Consent's Revenge." *Cultural Anthropology* 31 (3): 326–33.

Sloterdijk, Peter, 1988. *Critique of Cynical Reason*. Minneapolis: University of Minnesota Press.

Smith, Benjamin T. 2009. *Pistoleros and Popular Movements: The Politics of State Formation in Postrevolutionary Oaxaca*. Lincoln: University of Nebraska Press.

Smith, Benjamin T. 2012. "Heliodoro Charis Castro and the Soldiers of Juchitán: Indigenous Militarism, Local Rule, and the Mexican State." In *Forced Marches: Soldiers and Military Caciques in Modern Mexico*, edited by Ben Fallaw and Terry Rugeley, 110–35. Tucson: University of Arizona Press.

Smith, Derek A., Peter H. Herlihy, John H. Kelly, and Aida Ramos Viera. 2009. "The Certification and Privatization of Indigenous Lands in Mexico." *Journal of Latin American Geography* 8 (2): 175–207.

Smith, Sherry L., and Brian Frehner, eds. 2010. *Indians and Energy: Exploitation and Opportunity in the American Southwest*. Santa Fe: SAR Press.

Smith-Nonini, Sandy. 1998. "Health 'Anti-Reform' in El Salvador: Community Health NGOs and the State in the Neoliberal Era." *PoLAR* 21 (1): 99–113.

Star, Susan Leigh. 1999. "The Ethnography of Infrastructure." *American Behavioral Scientist* 43 (3): 377–91.

Star, Susan Leigh, and Karen Ruhleder. 1996. "Steps toward an Ecology of Infrastructure: Borderlands of Design and Access for Large Information Spaces." *Information Systems Research* 7 (1): 111–34.

Stengers, Isabelle. 2005. "The Cosmopolitical Proposal." In *Making Things Public*, edited by Bruno Latour and Peter Weibel, 994–1003. Cambridge, MA: MIT Press.

Stephen, Lynn. 1997. "The Zapatista Opening: The Movement for Indigenous Autonomy and State Discourses on Indigenous Rights in Mexico, 1970–1996." *Journal of Latin American Anthropology* 2: 2–41.

Stephen, Lynn. 2012. "Community and Indigenous Radio in Oaxaca: Testimony and Participatory Democracy." In *Radio Fields: Anthropology and Wireless Sound in the 21st Century*, edited by Lucas Bessire and Daniel Fisher, 124–41. New York: New York University Press.

Stephen, Lynn. 2013. *We Are the Face of Oaxaca: Testimony and Social Movements*. Durham, NC: Duke University Press.

Stern, Alexandra Minna. 2002. "From Mestizophilia to Biotypology: Racialization and Science in Mexico, 1920–1960." In *Race and Nation in Modern Latin America*, edited by Nancy P. Applebaum, Anne S. Macpherson, and Karin A. Rosenblatt, 187–210. Chapel Hill: University of North Carolina Press.

Stevenson, Mark. 2012. "Mexico Wind Farms Pit Indigenous against Corporations." *Associated Press*. October 31.

Strauss, Sarah, and Ben Orlove, eds. 2003 *Weather, Climate, Culture*. Oxford: Berg.

Strauss, Sarah, Thomas Love, and Stephanie Rupp, eds. 2013. *Cultures of Energy*. Walnut Creek, CA: Left Coast.

Sunder Rajan, Kaushik, ed. 2012. *Lively Capital: Biotechnologies, Ethics, and Governance in Global Markets*. Durham, NC: Duke University Press.

Sutherland, Keston. 2008. "Marx in Jargon." *World Picture* 1 (Spring). http://worldpicturejournal.com/WP_1.1/KSutherland.html. Accessed February 6, 2015.

Swyngedouw, Erik. 2009. "The Antinomies of the Postpolitical City: In Search of a Democratic Politics of Environmental Production." *International Journal of Urban and Regional Research* 33 (3): 601–20.

Szeman, Imre. 2007. "System Failure: Oil, Futurity and the Anticipation of Disaster." *South Atlantic Quarterly* 106 (4): 805–23.

Szeman, Imre. 2013. "What the Frack? Combustile Water and Other Late Capitalist Novelties." *Radical Philosophy* 177 (January/February). http://www.radicalphilosophy.com/commentary/what-the-frack.

Taussig, Michael. 1987. *Shamanism, Colonialism, and the Wild Man: A Study in Terror and Healing*. Chicago: University of Chicago Press.

Taylor, Analisa. 2009. *Indigeneity in the Mexican Cultural Imagination*. Tucson: University of Arizona Press.

Taylor, Bron. 2010. *Dark Green Religion: Nature Spirituality and the Planetary Future*. Berkeley: University of California Press.

Taylor, Maxwell D. 1972. *Swords and Ploughshares: A Memoir*. New York: Norton.

Taylor, William B. 1972. *Landlord and Peasant in Colonial Oaxaca*. Stanford, CA: Stanford University Press.

Terán, Victor. 2012. "La lengua de los binnizá." *Quadratín*, March 22.

Tiedje, Kristina. 2008. "Que sucede con PROCEDE? (What Is Happening with PROCEDE?) The End of Land Restitution in Rural Mexico." In *The Rights and Wrongs of Land Restitution: "Restoring What Was Ours,"* edited by Derick Fay and Deborah James, 209–34. New York: Routledge-Cavendish.

Tillman, Laura. 2017. "How Mexico's President Saw His Approval Rating Plummet to 17%." *Los Angeles Times*, March 1. http://www.latimes.com/world/mexico-americas/la-fg-mexico-president-20170301-story.html.

Treré, Emiliano. 2015. "Reclaiming, Proclaiming, and Maintaining Collective Identity in the #YoSoy132 Movement in Mexico: An Examination of Digital Frontstage and Backstage Activism through Social Media and Instant Messaging Platforms." *Information, Communication and Society* 18 (8): 901–15.

Trouillot, Michel-Rolph. 2001. "The Anthropology of the State in the Age of Globalization: Close Encounters of the Deceptive Kind." *Current Anthropology* 42 (1): 125–38.

Tsing, Anna Lowenhaupt. 2004. *Friction: An Ethnography of Global Connection.* Princeton, NJ: Princeton University Press.

Tsing, Anna Lowenhaupt. 2015. *The Mushroom at the End of the World: On the Possibility of Life in Capitalist Ruins.* Princeton, NJ: Princeton University Press.

Tsing, Anna Lowenhaupt, Heather Anne Swanson, Elaine Gan, and Nils Bubandt, eds. 2017. *Arts of Living on a Damaged Planet: Ghosts and Monsters of the Anthropocene.* Minneapolis: University of Minnesota Press.

Tucker, William P. 1957. *The Mexican Government Today.* Minneapolis: University of Minnesota Press.

Tutino, John. 1980. "Rebelión indígena en Tehuantepec." *Cuadernos Políticos* 24 (April–June): 89–101.

Tutino, John. 1993. "Ethnic Resistance: Juchitán in Mexican History." In *Zapotec Struggles: Histories, Politics and Representations from Juchitán, Qaxaca,* edited by Howard Campbell, Leigh Binford, Miguel Bartolomé, and Alicia Barabas, 41–61. Washington, DC: Smithsonian Institute.

UNIDO. 1975. "Lima Declaration and Plan of Action on Industrial Development and Co-Operation." Second General Conference of the United Nations Industrial Development Organization, Lima, Peru. March 12–26. https://www.unido.org/fileadmin/media/images/1975-Lima_Declaration_and_Plan_of_Action_on_Industrial_Development_and_Co-operation_26.3.1975.pdf.

Van Dun, Mirella. 2014. "Exploring Narco-Sovereignty/Violence: Analyzing Illegal Networks, Crime, Violence, and Legitimation in a Peruvian Cocaine Enclave (2003–2007)." *Journal of Contemporary Ethnography* 43 (4): 395–418.

Vannier, Christian N. 2010. "Audit Culture and Grassroots Participation in Rural Haitian Development." *PoLAR* 33 (2): 282–305.

Varela Guinot, Helena. 1993. *La oposición dentro del PRI y el cambio político en México (1982–1992): Crisis y transformación de un regimen autoritario.* Madrid: Editorial Centro de estudios avanzados en Ciencias Sociales.

Vázquez, Francisco Reveles, ed. 2003. *Partido Revolucionario Institucional: Crisis y refundación.* Mexico City: Ediciones Gernika.

Velázquez Becerril, César Arturo. 2010. "Intelectuales y Poder en el Porfiriato: Una Aproximación al Grupo de los Científicos, 1892–1911." *Fuentes Humanísticas* 41: 7–23.

Venn, Fiona. 2002. *The Oil Crisis.* New York: Routledge.

Veracini, Lorenzo. 2010. *Settler Colonialism: A Theoretical Overview.* London: Palgrave Macmillan.

Veracini, Lorenzo. 2015. *The Settler Colonial Present.* New York: Palgrave Macmillan.

Verdery, Katherine. 1996. *What Was Socialism and What Comes Next?* Princeton, NJ: Princeton University Press.

Von Schnitzler, Antina. 2013. "Traveling Technologies: Infrastructure, Ethical Regimes, and the Materiality of Politics in South Africa." *Cultural Anthropology* 28 (4): 670–93.

Von Tempsky, Gustav Ferdinand. 1858. *Mitla: A Narrative of Incidents and Personal Adventures on a Journey in Mexico, Guatemala, and Salvador in the Year 1853 to 1855.* London: Longman, Brown, Green, Longmans & Roberts.

Wacquant, Loïc. 2008. "The Militarization of Urban Marginality: Lessons from the Brazilian Metropolis." *International Political Sociology* 2: 56–74.

Wallerstein, Immanuel. 1974: *The Modern World-System,* vol. 1, *Capitalist Agriculture and the Origins of the European World-Economy in the Sixteenth Century.* New York: Academic.

Wallerstein, Immanuel. 1979: *The Capitalist World-Economy.* Cambridge: Cambridge University Press.

Webber, Jude. 2014. "Oil Companies Eye Mexico's 'Sweet Shop.'" *Financial Times,* September 26.

White, Leslie. 1943. "Energy and the Evolution of Culture." *American Anthropologist* 45 (3): 335–56.

White, Leslie. 1949. *The Science of Culture: A Study of Man and Civilization.* New York: Farrar, Straus and Giroux.

White, Leslie. 1959. *The Evolution of Culture: The Development of Civilization to the Fall of Rome.* New York: McGraw-Hill.

Wiest, Raymond E. 1978. *Rural Community Development in Mexico.* Winnipeg: University of Manitoba Anthropological Papers 21.

Wilhite, Harold. 2005. "Why Energy Needs Anthropology." *Anthropology Today* 21 (3): 1–3.

Wilson, Tamar Diana. 2010. "The Culture of Mexican Migration." *Critique of Anthropology* 30 (4): 399–420.

Winther, Tanja. 2008. *The Impact of Electricity: Development, Desires and Dilemmas.* New York: Berghahn.

Wionczek, Miguel S. 1965. "The State and the Electric-Power Industry in Mexico, 1895–1965." *Business History Review* 39 (4): 527–66.

Wolf, Eric R. 1956. "Aspects of Group Relations in a Complex Society: Mexico." *American Anthropologist* 58 (6): 1065–78.

Wolf, Eric R. 1982. *Europe and the People without History.* Berkeley: University of California Press.

Wolf, Eric R., and Edward C. Hansen. 1967. "Caudillo Politics: A Structural Analysis." *Comparative Studies in Society and History* 9: 168–79.

Zafra, Gloria, Jorge Hernández-Díaz, and Manuel Garza Zepeda. 2002. *Manifestaciones de la acción colectiva en Oaxaca.* Mexico City: Playa y Valdés.

Zeitlin, Judith Francis. 1989. "Ranchers and Indians on the Southern Isthmus of Tehuantepec: Economic Change and Indigenous Survival in Colonial Mexico." *Hispanic American Historical Review* 69 (1): 23–60.

Zeitlin, Judith Francis. 1994. "Precolumbian Barrio Organization in Tehuantepec, Mexico." In *Caciques and Their People*, edited by Joyce Marcus and Judith F. Zeitlin, 275–300. Ann Arbor: Museum of Anthropology, University of Michigan.

Zeitlin, Judith Francis. 2005. *Cultural Politics in Colonial Tehuantepec: Community and State among the Isthmus Zapotec, 1500–1750*. Stanford, CA: Stanford University Press.

Žižek, Slavoj. 1997. "Desire: Drive = Truth: Knowledge." *Umbr(a)* 1: 147–52.

Žižek, Slavoj. 1999. *The Ticklish Subject: The Absent Centre of Political Ontology*. London: Verso.

Žižek, Slavoj. 2002. *Revolution at the Gates: Žižek on Lenin—the 1917 Writings*. London: Verso.

Index

La Ventosa, 19, *65, 68, 70, 79, 82, 91,* 194;
aeolian politics in, 77–88, 93–94; *caciques*
in, 64–71, 149; private-public partnership
model, 24, 61; survey of residents,
88–94
Vestas, 75, 127–28
Villagómez Altamirano, Ramón, 139–41
Volkswagen, 77, 133
vulnerable state, 1, 22, 132

Wall Street Journal (newspaper), 129
Wal-Mart, 184
wind development: aeolian futures of, 194–98;
aeolian politics of, 22–23, 25, 28, 29, 30,
93–94, 125–26, 127, 130–31, 135–42, 159–64,
192, 194–98; areas with capacity, 139; CFE
on, 135–42; climate change and, 89, 93,
136; environmental impact of, 84; global
development of, 28–29, 42; grid voltage,
48–52; map, *51*; problems in, 107–12, 147–48,
149–50; resistance to, 103, 127–29, 155–56,
159–64, 173; visionaries in, 39–44. *See also*
autoabastecimiento; specific locations
World Bank, 105
World Wind Energy Institute network, 28

Yansa Group, 28–34, 127
Yansa-Ixtepec CIC, 23–24, 32–34, 42, 46, 54,
58–59, 104, 197; uprising, 34–39
#YoSoy132 (movement), 191

Zaachila, 166
Zapatistas, 41
Zapotec language (*diidxazá*), 40–41, 64,
67–68, 150, 160, 172, 220n4
Zapotec people (*binnizá*), 20, 22, 64, 68, 79,
92, 121, 127, 146, 165–66, 169, 170–73, 184, 187
Zárate, Luis, 112
Zedillo, Ernesto, 73
Zeitlin, Judith Francis, 168
Žižek, Slavoj, 2, 21
Zoques, 166